Climate Change and Agrometeorology

Latief Ahmad • Asim Biswas •
Jon Warland • Insha Anjum

Climate Change and Agrometeorology

Springer

Latief Ahmad
School of Environmental Sciences
University of Guelph
Ontario, ON, Canada

Asim Biswas
School of Environmental Sciences
University of Guelph
Ontario, ON, Canada

Jon Warland
School of Environmental Sciences
University of Guelph
Ontario, ON, Canada

Insha Anjum
Faculty of Horticulture
SKUAST-Kashmir
Jammu and Kashmir, India

ISBN 978-981-99-4862-8 ISBN 978-981-99-4863-5 (eBook)
https://doi.org/10.1007/978-981-99-4863-5

This Springer imprint is published by the registered company Springer Nature Singapore Pte Ltd.
The registered company address is: 152 Beach Road, #21-01/04 Gateway East, Singapore 189721, Singapore

Paper in this product is recyclable.

Preface

Meteorology is the study of the Earth's atmosphere and the variations in temperature and moisture patterns that produce different weather conditions. Agricultural meteorology or Agrometeorology is the applied branch of meteorology, which deals with the relationship between climate, weather, and life and growth of cultivated plants and animals.

There is hardly a phenomenon that has day-to-day impact on the human life and activities of agriculture than weather. Agriculture production is largely dependent on weather and climate despite advances in modern technology as a whole.

Agriculture meteorology is not only important in planning cropping systems, their monitoring but also disease and pest control. The extended benefits of it are discussed in this book. Agriculture as well as its allied services such as livestock are dependent upon weather for their comfort and food supplies. Individual elements of weather influence crop production and livestock in distinct ways.

In this book, we have made an attempt to make agromet familiar to the common masses including the farmers as well as students and teachers. The basics of agromet have been covered in a subtle and sophisticated manner that will be not so difficult to understand even for a person who isn't intimate with this subject.

This book discusses the scientific components of the greenhouse effect and how it affects climate change. It also explains why this topic is crucial and how certain actions can lessen the negative social, economic, and environmental effects of a changing climate.

Guelph, ON, Canada Latief Ahmad
Guelph, ON, Canada Asim Biswas
Guelph, ON, Canada Jon Warland
Jammu and Kashmir, India Insha Anjum

Contents

About the Authors

Latief Ahmad is a Senior Assistant Professor (Agrometeorology) at Sher-e-Kashmir University of Agricultural Sciences and Technology of Kashmir, India and is currently Visiting Scientist at the School of Environmental Sciences, Ontario Agricultural College, University of Guelph, Canada. He is having more than 12 years teaching experience in the field of Agrometeorology. The author has published more than 110 research articles in peer-reviewed international and national journals. He has authored 5 books, 10 book chapters, and 29 extension bulletins. He has attended more than 35 national and international conferences/training programs and is a life member of five scientific associations. He has received the best research proposal award from SERB-DST and is the recipient of several awards. He has also been a visiting scientist at different Universities in North America where he worked under world renowned scientists. His areas of interest include agrometeorology, climate change studies, crop simulation modeling, precision farming, water management, and crop science.

Asim Biswas is an Associate Professor and the Graduate Program Coordinator at the School of Environmental Sciences, University of Guelph, Canada, and a member of the Royal Society of Canada College of New Scholars. He is also a visiting Professor at three other universities. He is the current President of the Canadian Society of Soil Science and chair or vice-chair of several other divisions and communities at different professional societies. His research program on sustainable soil management is focused on increasing the productivity and resilience of our land-based agri-food production systems in an environmentally sustainable way while accounting for changing climate, economy, technology, and production methodologies. Integrating soil data, sensors, and technology with fundamental knowledge of soil and crop production practices, his pioneering research resulted in strategies that enable producers and policy makers to increase land-based agricultural productivity in an environmentally and financially sustainable manner. His research program focuses on three major areas: (a) development and adoption of new sensing technologies, mathematical models and systems to produce rapid and relatively inexpensive data revealing spatial heterogeneity and temporal dynamics of key soil characteristics; (b) calibration and validation of these emerging sensing technologies in the most economic and ergonomic manner to optimize information value of thematic soil maps; and (c) understanding soil processes for better adoption of precision agriculture through the development of decision support systems.

Jon Warland is the Associate Dean (Academic) for the Ontario Agricultural College. He did his B.S., Soil Science, Cornell University; M.Sc., Forest Meteorology, University of British Columbia; and Ph.D., Micrometeorology, University of Guelph. His current research examines weather impacts on crop production in Ontario, and development of new instrumentation and measures to determine heat budgets on vegetated surfaces. Jon's academic and research activities are enhanced by a strong interest in physics and its integration to more effectively address challenges in the natural environment. Jon's present research focus is on both the theory of turbulent dispersion within plant canopies, and the measurement of trace gas exchange over vegetated surfaces. These combine a range of theoretical, laboratory, and field work. He strongly believes that theory and experiments progress best when walking hand-in-hand. In the long term, he would like to tackle issues of scale interaction and complex terrain. This area is essential not only for fully developed models of climate change and more accurate long-range forecasting, but also for developing reliable adaptation strategies for possible climate change scenarios.

Insha Anjum is a Research Scholar, a young author. She has done her studies in Horticulture from SKUAST-K. Ms. Insha is a meritorious student and has participated in many national and international conferences having more than eight publications to her credit. She is a member of various committees constituted by the University.

Agricultural Meteorology: A Preview 1

Abstract

Agricultural meteorology is the study of climate and weather in relation with crop production or it is an interaction between agriculture and the environment. It comprises the correlation of climatic variables with crop and livestock production and soil management wherein all meteorological, hydrological, pedological, and biological factors impact the production of agriculture. It is an interdisciplinary science that forms a bridge between physical and biological sciences. This branch of science integrates knowledge from various scientific disciplines that impact the cropping patterns or where the cropping pattern can impact weather and climate. This chapter, apart from introduction, provides statistics regarding scope of agricultural meteorology and importance of agricultural meteorology. Besides, this chapter also deals with the atmospheric composition and aspects of weather and climate.

Keywords

Climate · Weather · Agriculture · Meteorology · Atmosphere · Scope · Importance

1.1 Agricultural Meteorology

Agricultural meteorology is the study of climate and weather in relation with crop production or it is an interaction between agriculture and the environment. It comprises the correlation of climatic variables with crop and livestock production and soil management wherein all meteorological, hydrological, pedological, and biological factors impact the production of agriculture. It is an interdisciplinary science that forms a bridge between physical and biological sciences.

L. Ahmad et al., *Climate Change and Agrometeorology*,
https://doi.org/10.1007/978-981-99-4863-5_1

1.2 Scope

Weather plays an important role in the development of agriculture. Increased weather variations have necessitated the introduction of the subject that can bridge the gap between changing weather variables and the sustenance of agriculture. Agricultural meteorology has gained more importance since the climate change scenario has started to appear. It assists farmers by elucidating the effects: where they can apply the concerned knowledge and information in agrometeorological applications and practices.

1.2.1 Assortment of Agricultural Variables

Agricultural meteorology is used to assess the climatic stability in a particular area necessary for crop production and economical gains. Climatic stability is governed by governing the climatic variables such as air temperature, precipitation, solar radiation, wind speed, and relative humidity, which determine the growth, development, and yield of the crop.

1.2.2 Crop Management

Crop management practices such as weed management, fertilizer application, plant protection, irrigation calendar, and harvesting can be carried out precisely on the basis of weather data. As a result of this, operational forecasts taken out by meteorologists are essential for the operations to work.

1.2.3 Stability in Production by Crop Planning

Crop planning can be done by carrying out strategic crop management skills based on suitable moisture requirements of crop such as effective precipitation and soil moisture conditions, with the goal of achieving crop stability and decreasing crop failure under severe agrometeorological conditions.

1.2.4 Crop Monitoring

The optimum growth and overall health and yield of crops can be kept under check by using various meteorological tools such as crop growth models and remote sensing.

1.2.5 Crop Modeling and Yield Climate Relationship

Prediction models or crop models developed for specific purposes such as crop growth, yield components, and seed yield helps in evaluating cultivar stability under long-term weather using current and past weather data.

1.2.6 Research in Crop–Climate Relationship

Agricultural meteorology helps to resolve the complexities of plant processes in relation to its required microclimate by helping to understand relationship between crop and its climate.

1.2.7 Climate Extremes

Crops can be protected from climatic extremities such as floods, droughts, hail, and windstorms by forecasting and warning the farmers in advance.

1.2.8 Analyzing Soil Moisture Stress

Soil moisture stress and drought can be determined by climatic water balance. The effects can be effectively mitigated by undertaking different protective procedures such as irrigation, mulching, application of anti-transparent, defoliation, and thinning.

1.2.9 Livestock Production

Livestock production is an integral part of agriculture. The livestock production, growth, and development are affected by weather conditions. The weather conditions are studied in agricultural meteorology and breeds can be selected according to the conditions or amiable conditions can be provided for existing breeds.

1.2.10 Soil Formation

Climate is an important factor in soil formation and development as the process of soil formation depends on climatic variables such as temperature, precipitation, humidity, and wind.

1.3 Importance and Practical Utility

Agricultural production can be greatly influenced by climatic variables where even from germination to production of fruit can be adversely affected by extreme climatic variables. The importance of agricultural meteorology in terms of agricultural production has been given as follows:

(a) Planning of cropping patterns or systems;
(b) Selection of sowing dates for increasing crop yield;
(c) Cost-effective farm operations such as ploughing, harrowing, and weeding;
(d) Decreasing losses of applied chemicals and fertilizers; helps to avoid fertilizer and chemical sprays when rain is expected;
(e) Judicious irrigation to crops;
(f) Well-organized harvesting of crops;
(g) Reduction or elimination in outbreak of pests and diseases;
(h) Efficient management of soils that are formed as a result of weather action;
(i) Managing weather abnormalities such as cyclones, heavy rainfall, floods, and drought can be effectively managed by weather forecasting; and
(j) Mitigation measures such as shelterbelts against cold and heat waves, effective environment protection, and so on.

1.4 Earth's Atmosphere

Atmosphere is a combination of two Greek words *atmos* meaning vapor and *sphaira* meaning globe, which got evolved in modern Latin as *atmosphera* and eventually into *atmosphere* in the mid-seventeenth century, and it refers to the layer of gases that surrounds the earth and is held in place by the earth's gravitational attraction (gravity). It is composed of a mixture of gases that are organized into layers, and this system together supports life on the earth. The earth's atmosphere supports life by creating pressure, thus allowing water to remain on the surface of the earth. It also absorbs ultraviolet radiations and creates greenhouse effect, thus reducing temperature extremes between day and night (diurnal temperature variation). While the atmosphere contains varied number of gases that are essential for the life to exist; however, the top four gases that make up 99.998% of the volume of clean dry air (unpolluted air that does not contain water vapor) are nitrogen, oxygen, carbon dioxide, and argon. The atmosphere also contains an appreciable amount of water vapor, as the atmosphere is rarely, if ever, completely dry.

In the dry composition of the atmosphere, nitrogen, by far, is the most common (78%). Nitrogen acts as the building block for protein formation; however, in its gaseous form it is unavailable to most of the living things. Nitrogen also prevents combustion process by diluting the oxygen that acts as an essential reactant in burning process. All living beings need oxygen; however, it is crucial for aerobic respiration. It is involved in many processes for the production of molecules that are essential for the sustenance of life and is also important for combustion and burning process. Carbon dioxide is an essential gas used by plants and other organisms to make sugar (food) through photosynthesis. This process is essential for other life as well because during photosynthesis, water molecules are split apart and their oxygen is released back to the atmosphere. Carbon dioxide also acts as a blanket that prevents the escape of heat into outer space. Argon is a nonreactive inert gas and is the most abundant noble gas in the earth's crust. It is used in incandescent bulbs and in energy efficient windows. It is also used by Scuba drivers for inflating dry suits and is importantly used to preserve priceless documents such as the original Declaration of Independence and the Constitution. Water vapor (water in a "gas" state) is usually present up to about 4% of the total volume depending on location. In the earth's desert regions (30° N/S) when dry winds blow, the water vapor contribution to the composition of the atmosphere will be near zero.

1.5 Weather and Climate

Weather is defined as a state or condition of atmosphere at a given place and a given time. It is basically daily variations or conditions of lower layers of the atmosphere. It pertains to smaller area and smaller duration of time that is part of a day or a complete day. Knowledge of weather and its elements is important to understand the role they have in growth and development of plant. Being a living entity, plant responds to the environment around it and the changes in that environment. To know about the role of weather in growth of a plant and its development, in various stages, one must be acquainted with the elements of weather that include atmospheric temperature, solar radiation, relative humidity, wind direction, wind speed, transpiration, precipitation, soil temperature, evaporation, and atmospheric pressure. Among these, solar radiation (sunlight), atmospheric temperature, and the amount of precipitation have a major role in anabolic and catabolic processes of plant metabolism. The changes in the weather are mostly observed in the place that is closest to the ground, known as troposphere. The factors responsible for causing these changes are temperature, humidity, direction, air pressure, and so on.

Climate is a broader term and is the average of the weather data for a place over a period of time, often 30 years. Climate information includes the statistical weather information that tells us about the normal weather as well as the range of weather extremes for a location. Different regions vary with different climates and depend on the factors of weathers such as humidity, temperature, precipitation, and rainfall that have occurred over a long period of time at that particular place. The main differences between weather and climate are explained in Table 1.1.

Table 1.1 Major differences between weather and climate

Comparison	Weather	Climate
Meaning	Weather is daily information of the changes in the atmospheric conditions in any area	Climate is statistical weather information that provides information about the average weather conditions of a particular place over a long period
Duration	The short-term atmospheric condition of any place is the weather, which may vary from time to time	The long-term average weather atmospheric condition of a place or country is the climate
Affected by	Weather is affected by temperature, pressure, humidity, cloudiness, wind, precipitation, rain, flooding, ice storms, and so on	Climate is the long-term observations of the atmospheric conditions at any location such as humidity, temperature, sunshine, and wind
It affects	Weather may affect the day-to-day occupation, and it may hamper transportation services, agriculture, and so on	Climate significantly affects agriculture, industries, and the livelihood of the people
Changes observed	The changes in the weather conditions can be observed very frequently	The changes in climate take a longer time to manifest
Studied by	Weather forecasting is monitored by the Meteorological Department of any particular location, and the study is known as Meteorology	The Climate Prediction Centre predicts climate, and its study is known as Climatology

Bibliography

Ahmad L, Kanth RH, Parvaze S, Mahdi SS (2017) Experimental agrometeorology: a practical manual. Springer, Cham, pp 83–87

Divanshi G (2019) Difference between weather and climate. https://biodifferences.com/difference-between-weather-and-climate.html

Zehnder C, Manoylov K, Mutiti S, Mutiti C, Vande Voort A, Bennett D (2015) Introduction to environmental science. In: Biological Sciences Open Textbooks. University System of Georgia, Atlanta, GA

Check for updates

Atmospheric Temperature

2

Abstract

Temperature is the thermodynamic condition that represents the hotness or coldness of the body. The value of the temperature is given by the heat flow from the hotter body to the one at low temperature until both of them reaches the state of equilibrium. Temperature in agriculture directly impacts the plant growth and survival. In meteorological perspective, temperature is measured for various parameters, including air temperature and soil temperature in relation to vegetation sustenance. In this chapter, all of the temperature variables that impact the vegetation growth, as well as daily and seasonal temperature variations, have been discussed in detail.

Keywords

Air temperature · Soil temperature · Temperature inversion · Lapse rate · Energy balance

2.1 Temperature

Temperature is defined as "a physical quantity characterizing the mean random motion of molecules in a physical body" (WMO 1992). At a specific time, a body is said to be at specific thermodynamic condition, referred to as temperature, which is the measure of hotness or coldness of the body. The value of temperature is determined by net flow of heat that gets spontaneously transferred from one body to another, i.e., from a hotter body (one at higher temperature) to a colder body (one at lower temperature). In meteorological perspective, temperature is measured for various parameters such as air temperature, soil temperature, ground temperature, grass minimum temperature, and seawater temperature.

L. Ahmad et al., *Climate Change and Agrometeorology*,
https://doi.org/10.1007/978-981-99-4863-5_2

2.2 Units of Temperature

The temperature is usually expressed in degrees Celsius (°C) or degrees Fahrenheit (°F). The SI unit of temperature is Kelvin (K).

2.2.1 Temperature in Context to Agrometeorology

Air temperature is the measure of hotness or coldness of air surrounding an individual at a specific time and at specific place. It is the most commonly measured weather parameter. WMO (1992) defines air temperature as "the temperature indicated by a thermometer exposed to the air in a place sheltered from direct solar radiation." Atmospheric temperature has a significant influence on growth and development of plants and crops. It is a vital environmental factor that determines the distribution of plants. Air temperature affects all plants ranging from physiological processes to phenotype development. It also affects enzymatic-catalyzed reactions and cell membrane transport. Air temperature also regulates transpiration, respiration, and specific compound volatilization. Warmer temperature enhances the rate of phenological development. It also influences the water requirement of crops.

Soil temperature, on the other hand, is very crucial for the chemical and biological activities to take place. The sown seeds, plant roots, and microorganisms that inhabit the soil are all influenced by soil temperature. The physiochemical and life processes are directly affected by the temperature of the soil.

Importance of Temperature on Crops

- *Temperature governs root and shoot ratio of plant:*
  Air temperature influences the proper development of root and shoot of a plant. High air temperature reduces plant growth, which in turn reduces the root development. Soil temperature is more crucial as it damages the roots, resulting in lesser growth of shoot.
- *Temperature influences the physiological processes:*
  Plant temperature directly influences the function of the physiological processes. Soil temperature governs the uptake of water and nutrients needed for photosynthesis. High air temperature tends to speed up the processes by increasing the reaction rates, leading to increased phenological development.
- *Temperature determines the rate of crop development:*
  Temperature determines the rate of growth and development of the crop, and ultimately the yield. Soil temperature influences the germination of seeds and development of roots. Air temperature, on the other hand, determines the phenological stage of the plant. Temperature affects germination by three ways, i.e., moisture, hormone production, and enzyme activity. All these are directly or indirectly affected by temperature.
- *Temperature plays a vital role in mineralization:*
  Temperature plays a vital role in mineralization by the conversion of organically bound nitrogen into inorganic forms of ammonium and nitrate. Soil microbes are

responsible for nitrogen-containing organic matter into inorganic soluble forms, making it highly temperature and season dependent.

- *Temperature controls soil microbial activities:*
  Temperature is an important factor in regulating microbial activity and shaping the soil microbial community. Fungal and bacterial growth rates showed optimum temperatures around 25–30 °C, with lower values seen at higher temperatures.
- *Temperature affects the speed of reactions:*
  Temperature affects the speed of chemical reactions. Warm temperatures speed up reactions, while cold temperatures slow them down. Soil temperature affects the breakdown of parent material and regulates the activity of microorganisms. Both are important for adding and returning nutrients to soil.
- *Temperature influences the soil structure:*
  Increase in soil temperature causes temperature-induced dehydration of 2:1 clay minerals in the soil, leading to strong interactions among the clay particles, which in turn yield less clay and more silt-sized particles in the soil.

2.3 Factors Affecting Temperature

Heat at ground surface is propagated downward in the form of waves. The amplitude deceases with depth. Both meteorological and soil factors contribute in bringing about changes of soil temperature.

2.3.1 Meteorological Factors

2.3.1.1 Solar Radiation

The amount of solar radiation available at any given location and point of time is directly proportional to temperature. Air temperature is directly related to the global solar radiation; hence, more the solar radiation, the more will be the temperature. Down surging radiation causes heating of the earth's surface due to direct sunlight absorption and also the back radiation from the atmosphere, referred to as greenhouse effect. Solar radiation is responsible for heating up both the air and soil. Some part of the net solar radiation is responsible for evapotranspiration and heating of air (latent and sensible heat fluxes), and a significant amount of solar radiation is used to heat the soil conditional to the soil surface, referred to as ground heat flux. A large amount of heat to the soil is contributed by sky radiations in areas where the sun's rays must infiltrate the earth's atmosphere diagonally.

2.3.1.2 Wind

Wind favors the conduction process from atmosphere, which is essential to heat up the soil. The mountain and valley winds control the soil temperature. On a calm night, the maximum surface cooling can take place. However, on a windy night, some warmer air is mixed downward to the surface, which prevents the temperatures from dropping as quickly as they would on a clear night.

2.3.1.3 Evaporation and Condensation

The energy from solar radiation evaporates the water from soil, thereby rendering it unavailable for heating up of the soil. Hence, the more the evaporation, the greater will be the soil cooled and the lower will be its temperature. On the other hand, during condensation, when the water from the atmosphere is taken back by the soil and hold it there, it generates heat by the process of freezing, thereby increasing the temperature of the soil.

2.3.1.4 Rainfall

Temperatures can decrease during a rainfall event for many reasons, including evaporative cooling or movement of cold air caused by the rain. Depending on the humidity value in atmosphere, the rain can either decrease the atmospheric temperature or increase the atmospheric temperature. When the rain is produced by surplus humidity, it actually heats up the atmosphere by latent heat of vaporization, which is lost when water blends into liquid. Conversely, if the air is dry, the rain evaporates back into the atmosphere (endothermic reaction) and hence causes evaporative cooling of the atmosphere.

2.3.2 Soil Factors

2.3.2.1 Aspect and Slope

The aspect of a slope can make very significant influences on its local climate (microclimate). For example, a west-facing slope will be warmer than a sheltered east-facing slope because the sun's rays are in the west at the hottest time of day in the afternoon. Similarly, in the northern hemisphere, a south-facing slope (more open to sunlight and warm winds) will therefore generally be warmer and drier due to higher levels of evapotranspiration than a north-facing slope.

In the Himalayas, this effect can be seen to an extreme degree, with south-facing slopes being warm, wet, and forested, and north-facing slopes cold and dry but much more heavily glaciated. The southern slopes are usually warmer than the northern slopes in middle and high latitudes. The reason is that the southern slopes receive more insulation per unit area than the northern slopes where the slope is less exposed to the sun. The soil also affects the air temperature by radiation, particularly near the surface of the soil, and is responsible not only for entire diurnal increase in temperature but also for the conventional currents of the air.

2.3.2.2 Soil Texture

Texture specifies the comparative content of particles of various sizes such as sand, silt, and clay in the soil. Texture influences the ease with which the soil either gains or losses the heat, thus affecting soil temperature. For example, sandy soils warm up more rapidly than clay soils; the reason being lower heat capacity and poor thermal conductivity. Hence, the energy it receives is mainly focused in a thin layer, resulting in an unusual rise in temperature. Similarly, the radiation cooling at night is greater in light soils (sand) than in heavy soils (clay). In addition, the decrease in range of depth is more rapid in light soils than in heavy soils when they are dry, but slower when wet.

The roughness and smoothness of the soil particles also influence the temperature gain or loss. Since the particles with rough surface absorbs more solar radiation than the smooth ones, the soil having more rough texture will be warmer than the soil having smooth texture.

2.3.2.3 Tillage, Tilth, and Mulching

Crop residue remaining on the soil surface in conservation tillage systems can decrease the rate of soil temperature change because surface residue both increases the reflection of incident solar radiation and acts as an insulating barrier between the soil surface and the warmer (or colder) atmospheric air above. Mulching by loosening the top soil reduces the heat flow and overall loss of heat by creating a barrier between the surface and the subsoil. The soil mulch has a greater surface contact than the undisturbed soil, and there is no capillary linking with the underground moist layers. Hence, the cultivated soil dries up quickly by evaporation, but the moisture in the sub-soil beneath the dry mulch is conserved. The air 2–3 cm above the tilled soil is often hotter (10 °C or above) than that of an untilled soil. However, loosened ground is more liable to frost than the untilled land. In general, soil warms up faster than air. The diurnal temperature wave of the cultivated soil has much larger amplitude than that of the uncultivated soil.

2.3.2.4 Organic Matter

Organic matter increases the water-holding capacity of the soil. It also contributes to the dark color of the soil. These two soil properties increase heat absorption, thereby increasing the soil temperature. Due to this dark color of soil, the absorptivity of the soil also increases; hence, the fraction of reflected radiation also reduces. Organic matter, however, has low thermal conductivity, and hence the flow of heat from sub-soil is reduced by radiation during nighttime due to addition of organic matter.

2.3.2.5 Soil Moisture

An increase in moisture content decreases the soil temperature differences between daytime and nighttime, which provides protection to the plant root system against sharp and sudden changes of soil temperature. It is also found that the absorption of solar energy increases as the moisture content increases, which results in a higher heat storage capacity at higher moisture content. Moisture is a good conductor of heat; hence, there will be more uniform temperature throughout the depth of the soil. Moisture at the soil surface also cools the soil through evaporation; hence, moist soil at surface is cooler than dry soil.

2.3.2.6 Daily Variations of Soil Temperature

These variations occur at the surface of the soil due to solar radiations that fall on soil surface, resulting in a temperature difference between soil surface and subsoil. On a clear sunny day, a bare soil surface is hotter than the air temperature. During daytime, heat flows from surface soil to subsoil, but at nighttime the reverse occurs. So, the profile goes through different temperature variations, called diurnal variations of soil temperature. The amplitude of diurnal wave of soil temperature decreases rapidly with depth and is smaller at higher depth. At around 20 cm in the soil, the temperature in the ground reaches a peak after the surface reaches its maximum due to the longer time the heat takes to penetrate the soil. The rate of penetration of heat wave within the soil takes around 3 h to reach 10 cm depth. At 5 cm depth the change exceeds 10 °C, at 20 cm the change is minor, and at 80 cm diurnal changes are practically insignificant. Sandy soils have greater diurnal variations than clayey soils. On a clear sunny day, the temperature fluctuations are more significant in comparison to that of a cloudy day, which is due to the fact that on a cloudy day, the soils become wet, which in turn increases the overall heat capacity of the soil. The cooling period and warming period of the soil surface are different. The cooling period of the daily cycle of the soil surface temperature is almost double than the warming period, which leads to greater fluctuations in diurnal variations of soil temperature. Undesired daily temperature variations can be minimized by scheduling irrigation.

2.3.2.7 Seasonal Variations of Soil Temperature

Seasonal variations occur much deeper into the soil. Variations occur with change in temperature during different seasons. It depends on the climatic conditions of different places. In winter the soil surfaces are at minimum temperature and the soil is progressively warmer at lower depths. Therefore, the direction of net flow of heat is upwards. In spring season the soil surface begins to warm up depending upon the net solar radiation that falls on the soil surface. Therefore, the heat flow now reverses in the other direction, i.e., from soil surface to subsoil. The whole soil profile attains maximum temperature at the end of May or June. In monsoon, the variations are very little due to increased heat capacity of the soil. However, minor net heat flow occurs from surface soil to subsoil. After the commencement of October, movement of heat takes place from bottom layers to upper surface and the cycle continues.

2.3.3 Lapse Rate

Lapse rate of change in temperature observed while moving upward through the earth's atmosphere. The lapse rate is considered "positive" when the temperature decreases with elevation, "zero" when the temperature is constant with elevation, and "negative" when the temperature increases with elevation (temperature inversion). The lapse rate of non-rising air – commonly referred to as the normal, or environmental, lapse rate – is highly variable, being affected by radiation, convection, and condensation; it averages about 6.5 °C per kilometer (18.8 °F per mile) in the lower atmosphere (troposphere).

2.3.4 Temperature Inversion

Temperature inversion, also called thermal inversion, is a reversal of the normal behavior of temperature in the troposphere (the region of the atmosphere nearest earth's surface) in which a layer of cool air at the surface is overlain by a layer of warmer air. (In normal conditions, air temperature usually decreases with height.) Inversions play an important role in determining cloud forms, precipitation, and visibility. An inversion acts as a cap on the upward movement of air from the layers below. As a result, convection produced by the heating of air from below is limited to levels below the inversion. Diffusion of dust, smoke, and other air pollutants is likewise limited. Inversions also affect diurnal variations in air temperature. The principal heating of air during the daytime is produced by its contact with a land surface that has been heated by the sun's radiation. Heat from the ground is communicated to the air by conduction and convection.

Since an inversion will usually control the upper level to which heat is carried by convection, only a shallow layer of air will be heated if the inversion is low and large, and the rise in temperature will be great. There are four types of inversions: ground, turbulence, subsidence, and frontal. A ground inversion develops when air is cooled by contact with a colder surface until it becomes cooler than the overlying atmosphere; this occurs most often on clear nights when the ground cools off rapidly by radiation. If the temperature of surface air drops below its dew point, fog may result. Topography greatly affects the magnitude of ground inversions. If the land is rolling or hilly, the cold air formed on the higher land surfaces tends to drain into the hollows, producing a larger and thicker inversion above low ground and little or none above higher elevations. A turbulence inversion often forms when quiescent air overlies turbulent air. Within the turbulent layer, vertical mixing carries heat downward and cools the upper part of the layer. The unmixed air above is not cooled and eventually is warmer than the air below; an inversion then exists.

A subsidence inversion develops when a widespread layer of air descends. The layer is compressed and heated by the resulting increase in atmospheric pressure and, as a result, the lapse rate of temperature is reduced. If the air mass sinks low enough, the air at higher altitudes becomes warmer than at lower altitudes, producing a temperature inversion. Subsidence inversions are common over the northern continents in winter and over the subtropical oceans; these regions generally have subsiding air because they are located under large high-pressure centers. A frontal inversion occurs when a cold air mass undercuts a warm air mass and lifts it aloft; the front between the two air masses then has warm air above and cold air below. This type of inversion has a considerable slope, whereas other inversions are nearly horizontal. In addition, humidity may be high and clouds may be present immediately above it.

2.3.5 Energy Balance of Earth

The earth–atmosphere energy balance is the balance between incoming energy from the sun and outgoing energy from the earth. Energy released from the sun is emitted as short-wave light and ultraviolet energy. When it reaches the earth, some part is reflected back to space by clouds, some part is absorbed by the atmosphere, and some part is absorbed at the earth's surface. However, since the earth is much cooler than the sun, its radiating energy is much weaker (long wavelength) infrared energy. The earth–atmosphere energy balance is achieved as the energy received from the sun balances the energy lost by the earth back into space. In this way, the earth maintains a stable average temperature and therefore a stable climate. Sun is the ultimate incoming source of energy and this energy comes from the sun as short-wave radiations. These short-wave radiations at the top of the earth's atmosphere are reflected back to space by many objects either in the form of short-wave radiations or long-wave radiations, referred to as outgoing energy. Short-wave radiations are sent back to space by clouds present in the atmosphere and by surface of the earth.

However, the outgoing energy in the form of long-wave radiations is sent back to the space by atmosphere, clouds, and earth's atmosphere. After entering the atmosphere, the incoming source of energy is now the absorbed energy of solar radiation. These can be short wave (e.g., gases in the atmosphere and clouds) or long wave (e.g., from the earth's surface or convective currents) (rising air warms the atmosphere). Condensation of water vapor also releases the heat into atmosphere. The outgoing energy in the earth's atmosphere is long radiation only, which are emitted by clouds, earth's surface, and gases of atmosphere.

At the surface of the earth, the incoming source of energy becomes the absorbed short-wave radiations from the sun and absorbed long-wave radiations from gases in atmosphere. Consequently, the outgoing energy source will be long-wave radiations from the surface of the earth, heat required for evaporation, and removal of heat by convection (rising warm air).

The absorption of infrared radiation trying to escape from the earth back to space is particularly important to the global energy balance. Energy absorption by the atmosphere stores more energy near its surface than it would if there was no atmosphere. The average surface temperature of the moon, which has no atmosphere, is 0 °F (−18 °C). By contrast, the average surface temperature of the earth is 59 °F (15 °C). This heating effect is called the greenhouse effect.

2.3.6 Vertical Profiles of Temperature

The temperature and pressure profiles of the atmosphere depend on the composition and constituents of the atmospheric gases, which differ and vary with geographical location, altitude, and season. Generally, the vertical structure of the earth's atmosphere has been described by the so-called standard atmospheres. The standard atmosphere used so far is the US Standard Atmosphere of 1976 (USSA 1976). Figure 2.1 shows the vertical temperature and pressure profiles, indicating the layers

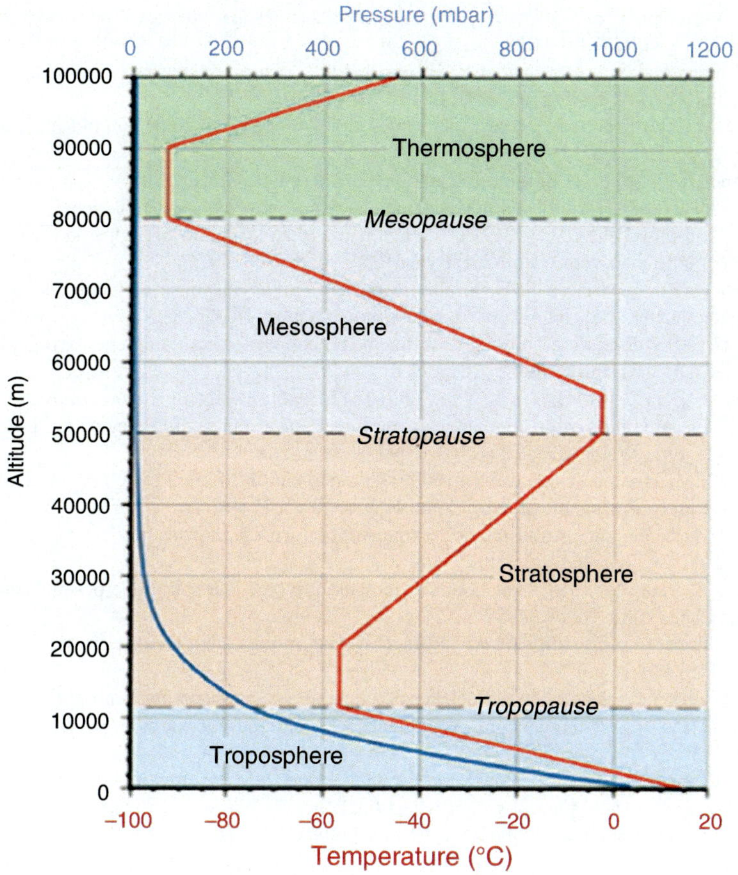

Fig. 2.1 Temperature and pressure profiles

of the lower atmosphere (0–11 km), called the troposphere; the lower-middle atmosphere (20–50 km), called the stratosphere; the upper-middle atmosphere (56–80 km), called the mesosphere; and the upper atmosphere (90–100 km), called the thermosphere. The turning points of the temperature profile are formed by intermediate layers: the tropopause (11–20 km), the stratopause (50–56 km), and the mesopause (80–90 km).

Bibliography

Abu-Hamdeh NH (2003) Thermal properties of soils as affected by density and water content. Biosyst Eng 86(1):97–102

Al-Kayssi AW, Al-Karaghouli AA, Hasson AM, Beker SA (1990) Influence of soil moisture content on soil temperature and heat storage under greenhouse conditions. J Agric Eng Res 45:241–252

Arocena JM, Opio C (2003) Prescribed fire-induced changes in properties of sub-boreal forest soils. Geoderma 113(1–2):1–16

Bao J, Sherwood SC, Alexander LV, Evans JP (2017) Future increases in extreme precipitation exceed observed scaling rates. Nat Clim Chang 7(2):128–132

Bever JD (1994) Feedback between plants and their soil communities in an old field community. Ecology 75(7):1965–1977

Britannica E (2016) Noet Edition. Encyclopædia Britannica, Chicago, IL

Britannica T, (2020) Editors of Encyclopaedia. Argon. Encyclopedia Britannica

Bui A, Johnson F, Wasko C (2019) The relationship of atmospheric air temperature and dew point temperature to extreme rainfall. Environ Res Lett 14(7):074025

Fang C, Smith P, Moncrieff JB, Smith JU (2005) Similar response of labile and resistant soil organic matter pools to changes in temperature. Nature 433(7021):57–59

Harde H (2013) Radiation and heat transfer in the atmosphere: a comprehensive approach on a molecular basis. Int J Atm Sci 2013:26

Li R, Hou X, Jia Z, Han Q, Ren X, Yang B (2013) Effects on soil temperature, moisture, and maize yield of cultivation with ridge and furrow mulching in the rainfed area of the Loess Plateau, China. Agric Water Manag 116:101–109

Moreno MM, Moreno A, Mancebo I (2009) Comparison of different mulch materials in a tomato (Solanum lycopersicum L.) crop. Span J Agric Res 2:454–464

National Oceanic and Atmospheric Administration (n.d.) https://www.weather.gov/jetstream/energy

Onwuka B, Mang B (2018) Effects of soil temperature on some soil properties and plant growth. Adv Plants Agric Res 8:34–37

Sciencedirect (n.d.-a) https://www.sciencedirect.com/topics/engineering/vertical-temperature-profile

Sciencedirect (n.d.-b) https://www.sciencedirect.com/topics/engineering/atmosphere-standard

Shreve F (1924) Soil temperature as influenced by altitude and slope exposure. Ecology 5(2): 128–136

Suying C, Xiying Z, Dong P, Hongyong S (2005) Effects of corn straw mulching on soil temperature and soil evaporation of winter wheat field. Trans Chin Soc Agric Eng 10:038

University of Illinois (n.d.) http://ww2010.atmos.uiuc.edu/(Gh)/wwhlpr/fcst_temps_winds.rxml

Wikipedia (n.d.) https://en.wikipedia.org/wiki/Aspect_(geography)#cite_ref-4

Check for updates

Atmospheric Pressure and Solar Radiation

3

Abstract

Atmospheric pressure or the barometric pressure is the pressure within the atmosphere of the earth and is defined as the force exerted by the air above, over the surface of the earth, as the gravity pulls it. Atmospheric pressure acts as an indicator of weather and is an important parameter in determining crop activity. Barometric pressure changes with change in altitude and is measured by an instrument called Barometer. In this chapter, various other topics have also been incorporated that are affected by atmospheric pressure such as solar radiation, depletion of solar radiation, thermal radiation, and net radiation.

Keywords

Atmospheric pressure · Solar radiation · Thermal radiation · Net radiation · Solar constant · Albedo

3.1 Atmospheric Pressure

Pressure is the force exerted on unit area. Atmospheric pressure, also known as barometric pressure on a given surface, is the force exerted per unit area by the weight of the earth's atmosphere above. It is the pressure within the atmosphere of the earth. Thus, the atmospheric pressure is equal to the weight of a vertical column of air above the earth's surface, extending to the outer limits of the atmosphere. Atmospheric pressure is measured by means of an instrument known as barometer.

L. Ahmad et al., *Climate Change and Agrometeorology*,
https://doi.org/10.1007/978-981-99-4863-5_3

3.2 Units of Measurement

The basic unit for atmospheric pressure measurements is the Pascal (Pa) (or newton per square meter). Meteorological pressure measurements are stated in terms of hectopascals (hPa), which is equal to 100 Pa. Previously, mb (millibar) was used, and 1 hPa is equal to 1 mb.

3.2.1 Factors Affecting Atmospheric Pressure

Atmospheric pressure in any region is affected by the following factors:

1. **Temperature:** Air pressure drops as the temperature cools down because of the gas molecules that tend to decrease the speed as they cool down. As the speed decreases, the collisions between the molecules and air decrease, resulting in a fall in air pressure. Air density acts as a correlation factor between temperature and pressure. As warmer air is less dense than cool air, molecules have more space to collide with greater force. However, in cooler air, the molecules are held closer together. The juxtaposition of air molecules results in greater collisions with less force and consequently lower air pressure.
2. **Altitude:** Atmospheric pressure decreases as the height from the sea level increases. It is mainly because of the fact that as one ascends, the number of gas molecules in the air decreases. As the air becomes less dense, there are fewer molecules per unit volume, resulting in fewer collisions. The air at higher elevations is also termed as "thin air" by meteorologists as it contains only lesser air molecules compared to that on the surface level. Thin air exerts less pressure than air at a lower altitude.
3. **Water vapor:** The addition of water vapor to the atmosphere decreases the atmospheric pressure. As humid air is less dense than dry air, both must be at same temperature; therefore, the addition of water vapor decreases the overall density and hence lowers its pressure. However, if there is temperature variation between the two, the pressure variant depends on temperature then as it results to more speed of molecules.
4. **Gravitational pull:** Air pressure has a proportional relationship with gravity. Due to gravitational pull, the atmosphere of the earth is intact, otherwise it would collapse. The air at the surface of the earth exerts more pressure than the air at higher elevations. In addition, the atmospheric pressure at poles is higher than at the equator at sea level, which is due to the fact that the equatorial radius is greater than the polar radius.
5. **Earth's rotation:** The earth's rotation does not affect the atmospheric pressure; however, it helps the air to circulate all over. If the earth did not rotate and remained stationary, the atmosphere would circulate only between the poles and the equator, i.e., the high-pressure zone and low-pressure zone, respectively, in a simple back-and-forth pattern. However, as the earth rotates, the circulating air is deflected, and this deflection is called the Coriolis Effect.

3.3 Variation of Pressure with Height

The pressure in the atmosphere at any level may be taken as the total weight of the air above a unit area at any elevation. At higher elevations, there are fewer air molecules above a given surface than at lower levels for the similar surface. For example, there are fewer molecules above the 50 km surface than are found above the 12 km surface, which is why the pressure is less at 50 km. Thus, it can be concluded that with increase in height, atmospheric pressure decreases considerably. Figure 3.1 shows that with increase in height there is significant decrease in atmospheric pressure. Since most of the molecules of the atmosphere are held close to the earth's surface by the force of gravity, air pressure decreases rapidly at first, then more slowly at higher levels.

Pressure decreases exponentially with altitude. More than half of the molecules of the atmosphere are located below an altitude of 5.5 km, atmospheric pressure decreases roughly 50% (to around 500 mb) within the lowest 5.5 km. Above 5.5 km, the pressure continues to decrease but at an increasingly slower rate. At 80 km altitude, the atmospheric pressure is down to 0.01 hPa, meaning that 99.999% of the atmosphere is below that altitude.

Near the earth's surface the pressure decreases with height at a rate of about 3.5 mb for every 30 m (100 ft.). However, over cold air the decrease in pressure can be much steeper because its density is greater than warmer air. At heights above 1500–3000 m (5000–10,000 ft.), the pressure is low enough to produce mountain sickness and severe physiological problems unless careful adjustments are commenced. Variations about these values on the surface of the earth are quite small; for example, the highest and lowest sea-level pressures ever recorded are

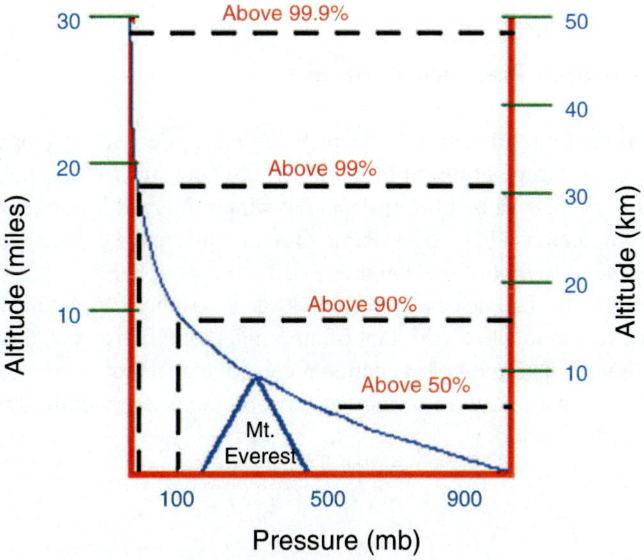

Fig 3.1 Variation of atmospheric pressure with height

32.01 in. (in the middle of Siberia) and 25.90 in. (in a typhoon in the South Pacific). However, these small variations in pressure that do exist largely determine the wind and storm patterns of the earth.

3.4 Diurnal and Seasonal Variation of Atmospheric Pressure

3.4.1 Diurnal Pressure Variation

Pressure shows a definite rhythm in the rise and fall at a specific region over the course of day. However, the areas closer to the sea show comparatively more variations than in the land areas, which is mainly because of the temperature differences between the two elements, i.e., sea and land, which creates the pressure difference and hence show more diurnal variation. The maxima and minima occur at approximately the same local time each day, keeping other related components constant.

The pressure variation varies greatly with the location. The effect of diurnal variations is more noticeable in tropics where incoming solar radiation is greatest but, as dynamic forcing is weak in these regions, absolute surface pressure changes are small. In case of mid-latitudes, surface pressure changes are dominated by dynamic forcing associated with polar front jet, which causes much greater short-term changes. However, the absolute air pressure variation in the tropics is much greater than mid-latitudes.

Radiation heating (air expansion) and radiation cooling (air contraction) are mainly responsible for diurnal variation in the air pressure. Equatorial regions absorb more heat than it loses while the polar regions give up more heat than they receive.

3.4.2 Seasonal Pressure Variation

Seasonal oscillations of atmospheric pressure are known to be closely associated with those of air temperature. In the summer, the land surface becomes warmer and atmospheric pressure falls. Moreover, in the winter the land is cooled faster than the sea; therefore, there is large atmospheric pressure difference between land and sea in winter compared to that of the summer when the atmospheric pressure difference is not so prominent. The seasonal dynamics of atmospheric pressure is sufficiently stable since it is associated with that of air temperature. However, the variation can be due to change in amount of insulation received in different seasons. In the tropical regions, the annual variation in atmospheric pressure is generally higher than the other parts of the world.

3.4.3 Relationship Between Atmospheric Pressure and Wind

Air pressure and winds have direct and positive correlation. The wind blowing and its velocity are dependent on pressure gradient. Pressure gradient is created when there is difference in air pressure. The more the intensity of the pressure difference, the more will be the pressure gradient. If the difference between two places is the same but the distance is different, then the smaller horizontal distance will show a steeper pressure gradient while the faster one will be gentler. Hence, it is the pressure difference per unit of distance travelled in a straight direction across isobars. Since atmospheric pressure is represented by isobars on the map that are much closer in the north-west direction than in the south-east direction where they are widely spread; winds blow much faster in the north-west direction than in the south-east direction, where the wind is very gentle.

3.5 Wind and its Types

The flow of gases and air on large scale is known as wind. It consists of bulk movement of air that flows on the surface of the earth. In meteorology, wind is the movement of air relative to the surface of the earth. Wind is the horizontal movement of air that moves around due to the forces acting on it. Winds occur due to horizontal and vertical gradients in atmospheric pressure, and hence play a vital role in controlling and determining the climate and weather of an area. Wind that is in a high-pressure system moves toward the area of low pressure, and this in turn causes wind. Temperature variations also affect the course of wind and its occurrence.

As atmospheric pressure largely controls the winds, its distribution is also affected accordingly by the atmospheric pressure. Near the earth's surface, winds generally flow around regions of relatively low and high pressure known as cyclones and anticyclones, respectively. In the lows of Northern Hemisphere, they rotate counterclockwise and vice versa in Southern Hemisphere. Similarly, wind systems rotate around the centers of highs in the opposite direction.

3.5.1 Common Wind Terminology

- **Gust**—It is a temporary and brief increase in wind speed that lasts for a few seconds, usually less than 20 s.
- **Squall**—A sudden increase of wind speed by at least three stages on the Beaufort scale, the speed rising to force 6 or more, and lasting for at least 1 min. It is common in fronts and can be identified by the approach of cumulus clouds.
- **Lull**—It is a noticeable decrease in wind speed lasting less than a few minutes.
- **Gale**—A gale is a very strong wind where the average surface wind ranges from 34 to 47 knots (kts).
- **Hurricane**—Average surface wind is greater than 63 kts.

- **Funnel effect**—When wind is forced to flow through narrow valley, it increases its speed considerably. This follows the Bernoulli theorem.
- **True Wind**—It is the actual meteorological wind that blows over the elements.
- **Apparent Wind**—Wind is felt while moving through the air. It may increase and decrease depending on the speed of the observer.
- **Storm**—A storm is any disturbed state of an environment or in the atmosphere of an astronomical body, especially affecting its surface and strongly implying severe weather. It may be marked by significant disruptions to normal conditions such as strong wind, tornadoes, hail, thunder and lightning (a thunderstorm), heavy precipitation (snowstorm, rainstorm), heavy freezing rain (ice storm), strong winds (tropical cyclone, windstorm), or wind transporting some substance through the atmosphere as in a dust storm, blizzard, and sandstorm.

3.6 Types of Winds

Wind blowing above the earth surface may be classified into five major types:

- Planetary winds
- Trade winds
- The westerlies
- Periodic winds
- Monsoon winds
- Land breeze
- Sea breeze
- Mountain and valley breeze
- Local winds

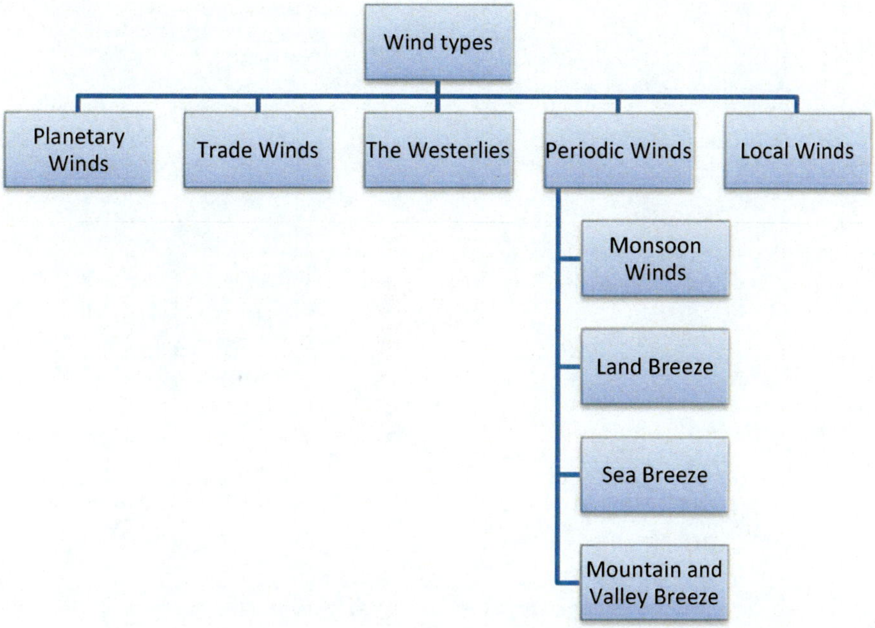

3.6.1 Planetary Winds

Planetary winds are the winds that blow regularly throughout the year in the lower strata of atmosphere. It is confined mainly in the regions of north-west and south-east, i.e., from higher pressure belts to lower pressure belts, respectively.

3.6.2 Trade Winds

These are air currents that blow from east to west, near the equator, and closer to the surface of the earth. Trade winds begin as warm, moist air rises from the equator into the atmosphere that cools and sinks down in the tropics of Northern Hemisphere and this phenomenon is known as Hadley cell, which has been shown in Fig. 3.2.

The trade winds blow toward the west partly because of how the earth rotates on its axis. The earth's rotation causes the circulating trade winds to partly shift toward the right in Northern Hemisphere and partly toward the left in Southern Hemisphere. This phenomenon is called the Coriolis effect as shown in Fig. 3.3. The trade winds can be found about 30° north and south of the equator. Right at the equator there is almost no wind at all – an area sometimes called the *doldrums*.

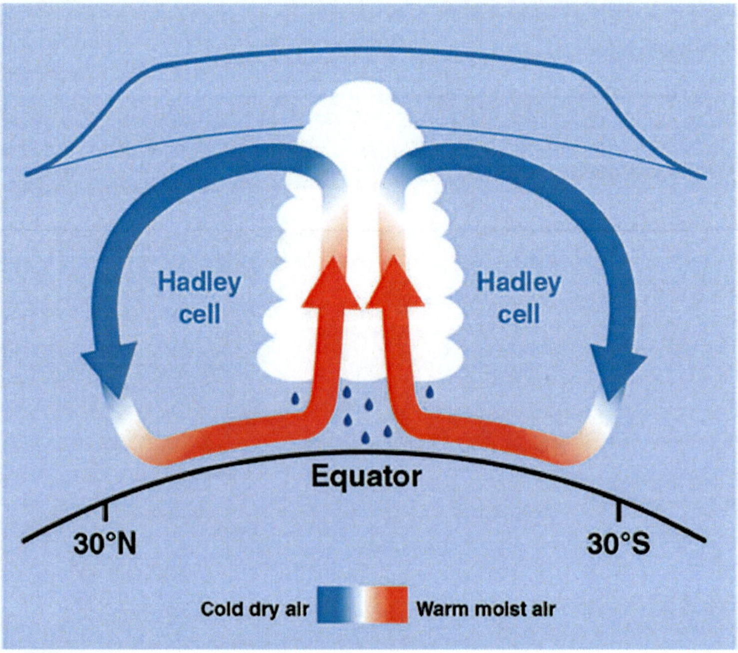

Fig. 3.2 Hadley cell phenomenon

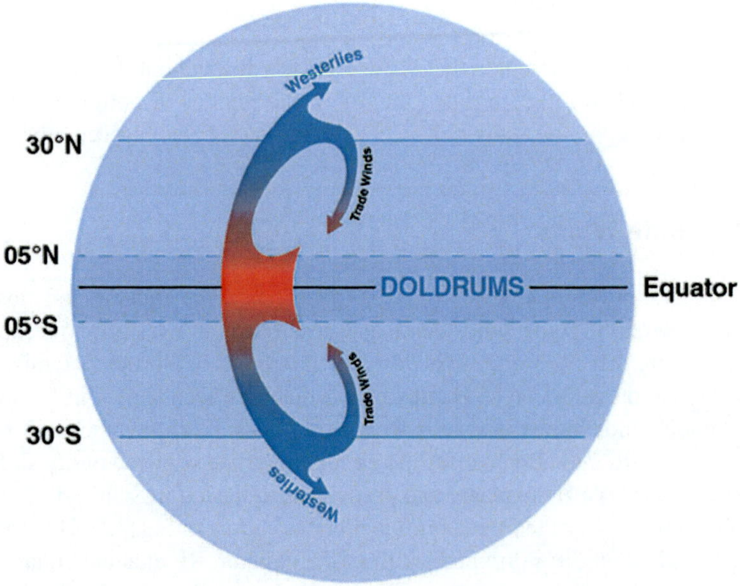

Fig. 3.3 Coriolis effect

3.6.3 The Westerlies

The *westerlies*, *anti-trades*, or *prevailing westerlies* are prevailing winds that blow from the west toward the east in the middle latitudes between 30° and 60° latitude. The name of these winds comes from the direction of their origin. The other entire winds blow from east to west, but the westerlies blow from west to east. They originate from the subtropical high-pressure areas in the horse latitudes present in both Northern Hemisphere and Southern Hemisphere and trend toward the poles. These are characterized by migratory cyclones and anticyclones travelling generally from west to east. Figure 3.4 shows the circulation patterns of the earth's surface that causes the disturbances leading to weather phenomenon.

Usually located poleward from approximately 35°N and S latitude, the westerlies extend much of the remainder of the distance to the poles and are developed to levels exceeding the height of the tropopause.

3.6.4 Periodic Winds

These are the winds that blow at regular intervals or in regular cycles and tend to change their direction periodically as there is change in season. These winds blow in a particular period of the year over a specific region and in a definite direction. These winds are a natural consequence of the localized difference in pressure and temperature.

Periodic winds are of following types:

- **Monsoon winds:** Monsoon winds are giant sea and land breezes produced by seasonal changes in circulation. The primary cause of monsoon winds is the temperature difference between land and sea. It is the same temperature difference

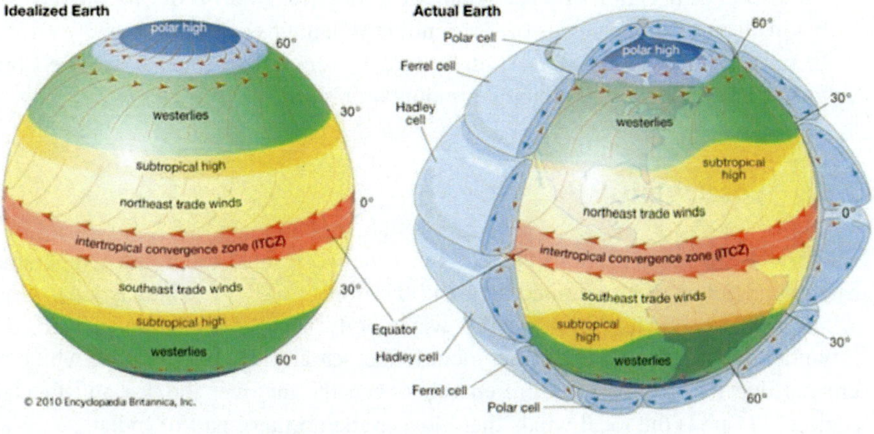

Fig. 3.4 Circulation patterns of the earth's surface

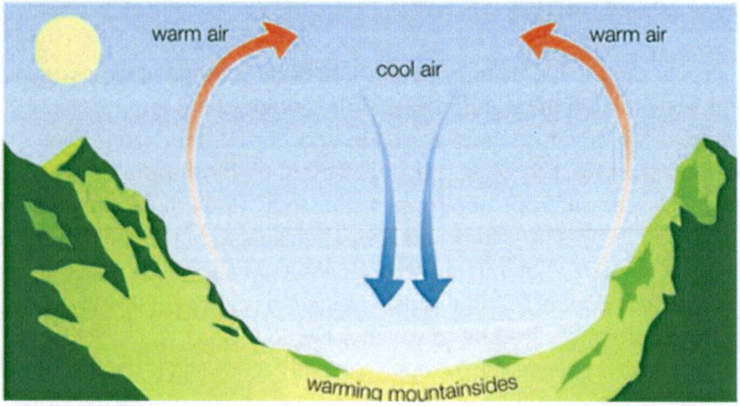

Fig. 3.5 Valley and mountain breeze

created by the Indian Ocean, Arabian Sea, Bay of Bengal, and Himalayan wall that forms the basis of monsoon in the Indian subcontinent.

- **Sea breeze:** These are the winds that blow from sea to land and carry some moisture. Seawater heats up more slowly as compared to land. As the temperature of the land rises, the air above it heats up by conduction. The density of warm air is less than the surrounding environment, because of which it rises, creating a vacuum in its place. The cooler air from the sea rushes in to fill the vacuum, creating a cool coastal breeze known as sea breeze.
- **Land breeze:** These winds blow from land to sea, carrying no moisture but dry and warm. At nighttime, the land cools off more quickly because of the low specific heat capacity compared to that of sea water, which creates a temperature difference between the temperature onshore and that of offshore. It is due to this temperature difference that again a pressure drop is created, resulting in the formation of a land breeze.
- **Mountain and valley breeze:** When the valley floor warms during the daytime, warm air rises up the slopes of surrounding mountains and hills to create a valley breeze as shown in Fig. 3.5. At nighttime, denser cool air slides down the slopes to settle in the valley, producing a mountain breeze.

3.6.5 Local Winds

These winds originate due to the difference between local relief and temperature. Such winds blow over a limited area and the impact of these winds is local. Local winds are also affected by a nearby water body such as an ocean, a lake, or a mountain range. Local winds also include the sea and the land breeze, which are created due to the pressure difference between the air over the sea and the land regions. “Loo” is the local winds that blow in the northern part of India.

On the basis of area covered, the winds may be classified as small-scale winds and large-scale winds.

3.6.6 Small-Scale Winds

One of the simplest examples of small-scale winds is the sea breeze. Seawater heats up more slowly as compared to land. As the temperature of the land rises, the air above it heats up by conduction. The density of warm air is less than the surrounding environment, because of which it rises, creating a vacuum in its place. The cooler air from the sea rushes in to fill the vacuum, creating a cool coastal breeze known as sea breeze.

3.6.7 Large-Scale Winds

A similar process takes place on a global scale. The sun's rays reach the earth's surface in polar regions at a much more slanted angle than at equatorial regions. This sets up a temperature difference between the hot equator and cold poles. So, the heated air rises at the equator, leading to low pressure, while the cold air sinks above the poles, leading to high pressure system. This sinking air then spreads out at the earth's surface, with some of it returning southward to the low pressure at the equator (known as trade winds), completing a circulation known as "global circulation patterns." However, this is complicated by the earth's rotation (known as the Carioles' effect).

Another portion of this air moves poleward and collides with the cold air flowing southward from the Arctic (or the Antarctic). As the warm air is less dense than the polar air, it tends to rise over it and this rising motion generates low-pressure systems that bring wind and rain to the shores. This part of the global circulation is known as the mid-latitude cell. The other important factor is that the Carioles' effect from the earth's rotation means that air does not flow directly from high to low pressure; however, it is deflected to the right in the Northern Hemisphere and vice versa in the Southern Hemisphere.

3.7 Measurement of Wind

Wind has two parameters that can be measured, i.e., speed and direction. Hence, two different devices are required to measure them, which are anemometer and wind vane.

3.7.1 Anemometers

These instruments are a common weather station instruments that are used to measure the speed of the wind. Cup anemometer, hot wire anemometer, windmill anemometer, sonic anemometer, and Laser Doppler anemometer are the different types of anemometers. Figure 3.6 shows the visual image of cup anemometer that measures the speed of wind.

3.7.2 Wind Vanes

The instrument used for the measurement of wind is wind vane as shown in Fig. 3.7. The direction from which the wind blows is known as the windward side, and toward which it bows is known as leeward side.

3.7.3 Daily and Seasonal Variation of Wind Speed

The winds represent air in motion, and the primary cause of all winds is regional differences in temperature and pressure. When these pressure differences persist for several hours, the rotation of the earth modifies the direction of motion till the winds blow along lines of equal pressure. Wind direction and speed are modified frequently

Fig. 3.6 Cup anemometer

Fig. 3.7 Wind vane

due to seasonal variation in solar radiation and differential heating of the earth's surface.

3.8 Land and Sea Breeze

3.8.1 Sea Breeze

It is a local wind system characterized by wind flow from sea to land during daytime. When the temperature rises, the sun's rays heat up the ground quickly, while as the sea surface has a greater capacity to absorb the sun's rays, it is more difficult to warm up. This leads to a temperature difference between the warm land and the cooler sea.

As the land heats up, it warms the air above it by the process of conduction. The warmer air becomes less dense than surrounding cooler air and hence begins to rise. The rising air creates a vacuum and thus results in lower pressure system over the land surface. The air over the sea, however, remains cooler and denser, so the pressure is higher than inland. This pressure difference creates a movement of air from the sea to inland surface in order to equalize this difference, as seen in Fig. 3.8, and the process is known as sea breeze.

Since the surface flow of the sea breeze sinks over land surfaces, a region of low-level air convergence is produced. Such convergence often leads to upward movement of air, which finally culminates in cloud formation. Such clouds may produce showers that occur over land in the afternoon. The existence and intensity of the sea breeze depends strongly on seasonal and latitudinal factors as well as on the time of day. On many tropical and subtropical coastal regions, the sea breeze is a regular phenomenon throughout the year, while in cooler regions the sea breeze is a common feature during spring and summer when the temperature difference between land and sea is at its maximum.

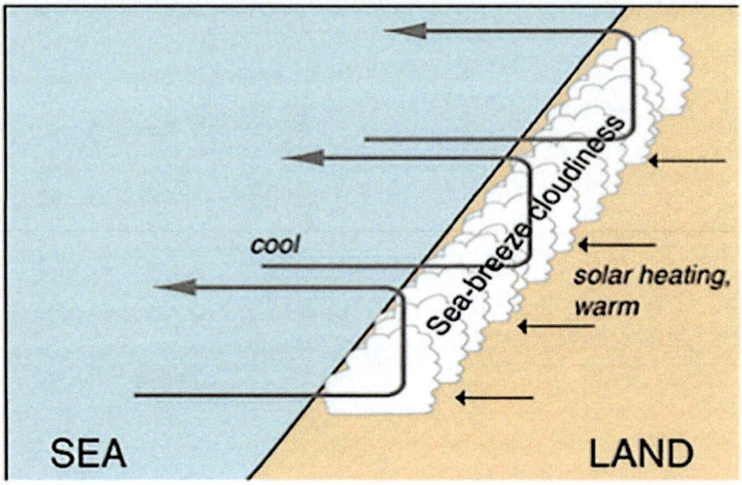

Fig. 3.8 Process of sea breeze

Those who live within 30–40 km (about 19–25 miles) of the coastline often experience the cooler, 10–20 km (about 6–12 mile) per-hour winds of the sea breeze on a sunny afternoon, only to have it turn into a sultry land breeze late at night.

3.8.2 Land Breeze

It is a local wind system characterized by a flow of air from land to water, especially at night. Land breezes alternate with sea breezes along coastlines adjacent to large bodies of water. Both are induced by differences that occur between the heating or cooling of the water surface and the adjacent land surface. The land breeze is typically shallower than the sea breeze since the cooling of the atmosphere over land is confined to a shallower layer at night than the heating of the air during the daytime. Since the surface flow of the land breeze terminates over water, a region of low-level air convergence is produced. Locally, such convergence often induces the upward movement of air, resulting in the development of clouds. Therefore, it is not uncommon to see clouds lying off the coast at night, which are later dissipated by the daytime sea breeze.

Land breezes are weaker than sea breezes but not because of the difference in heating. Daytime heating and nighttime cooling occur at about the same rate, so the potential for both land and sea breezes being the same strength exists. However, at nighttime, the cooling ground inhibits vertical motion, which in turn weakens the land breeze circulation.

Nighttime cooling also produces a shallower change in temperature, so land breeze circulation is shallower, and terrain, vegetation, and buildings inhibit the flow of air from land to water.

3.9 Effect of Wind on Crop Plants

The direction of wind and its velocity have significant influence on crop growth. Wind has both negative and positive impacts on plant growth and development, which are mentioned as follows:

- **Direct mechanical effects of wind:**
  The wind force can tear off leaves from the plant. Depending on the velocity and the growth stage of plant, the plant can either suffer mechanical tear off or they can be even knocked out from the soil. Dense plant canopies may suffer abrasion due to constant rubbing. Wind can also cause leaf tearing in many species, which can lead to temporary increase in water loss, and in addition, leaf folding in grasses can damage the cuticle and may lead to severe water loss and even wilting and death.

  Soil particles lifted into suspension by the wind have the potential to scrape and damage plant tissue. Recovery, and hence final yield, is dependent on the growth stage and soil/plant moisture status when the damage occurred, the particular species and variety, as well as the preceding and following weather.
- **Physiological impacts:**
  Wind and increased air movement around the plant result in higher transpiration rate, which is mainly due to the fact that the wind around the plant removes the saturated moist air away that creates a vacuum and thus the place is filled by dried air. It is this dried air that is responsible for increased transpiration from the plant. Since the transpiration is more, the water below the root zone cannot suffice the loss that results in wilting, and then permanent wilting if the plant is not provided additional source of water or irrigation, ultimately result in death of the plant. Hot winds also hasten the drying of the plants by replacing humid air by dry air in the intercellular spaces. For example, rice crop during June–July months shows tip drying. In addition, the hot and dry winds desiccate out the blossoms, thus hampering the overall fruit set.

  Moderate wind speeds can alter transpiration rates, indirectly affecting photosynthesis via changes in stomatal conductance and leaf temperature but this would be dependent on the local environmental conditions. For example, in a hot environment, increases in transpiration and concurrent decreases in leaf temperature could be beneficial, assuming water is not a constraint. However, under other less favorable conditions, the impact of moderate wind speeds on boundary layers alone may not affect leaf photosynthesis rates significantly or could negatively affect them.

  Wind directly influences photosynthesis and gaseous exchanges at the leaf scale as well as leaf temperature. Wind also modifies light shedding inside the plant canopy through the displacement or flutter of the outer leaves, thereby modifying photosynthesis.
- **Cause lodging of crops:**

Fig. 3.9 (**a**) Root lodging of wheat in a field that has become saturated by rain. (**b**) Brazier buckling or "brackling" in the stems of mature barley that have developed heavy seed heads

Many crops are particularly susceptible to wind damage by lodging. These include wheat, oats, barley, maize, grain sorghum, soybean, tomato, tobacco, rice, alfalfa, millet, sunflowers, and oilseed rape as shown in Fig. 3.9a. Lodging can take the form of stem lodging due to the bending or breaking of the lower culm internode or root lodging due to failure of root anchorage. Stem lodging is a consequence of the strength of the culm not being sufficient to prevent bending, leading to Brazier buckling from the combined moment of the wind loading on the leaves and seed heads of the crop plus the overhanging weight of the displaced top of the plant as seen in Fig. 3.9b. Winds also cause soil erosion that results in overexposing of roots of the plants, thus making them susceptible for lodging.

- **Causes economic yield loss:**
  Lodging reduces yields in many cereal crops and oilseed rape, particularly in high-yielding areas both through a reduction in the number of grains per m^2 and the average weight of the grain. The yield loss in wheat depends very much on the stage of development at which it occurs. Yield losses in wheat, rice, and maize result from inefficient use of light resulting from compression of the canopy. The greatest yield losses often occur following lodging at flowering or early grain filling. Lodging also reduces bread-making quality in wheat.
- **Diseases and secondary infection:**
  Winds open up the wounds by tearing off the plant part, thus opening the pathway for the diseases to occur. Soil particles that are lifted by wind also serve as the source of inoculum for the primary infection. If the fruit gets damaged by winds, it can invite many other secondary parasites, thus decreasing the overall yield. Winds can also pave way to disease spread by transporting the spores from one plant to another or from one area to another.

The positive impacts of winds are as follows:

- It helps in pollination and dispersal of seeds. Wind plays an effective role in pollination where wind is the only mode of pollination, i.e., anemophily such as wheat, maize, oat, and rice.
- Wind plays a significant role in determining and controlling the weather of an area. Winds greatly affect the amount of precipitation. The prevailing winds bring moisture from the ocean onto the continents and thus lead to precipitation in the particular area. The amount of precipitation in an area depends on the amount of moisture the winds carry.
- Atmospheric inversions can be prevented by moderate to strong winds. Strong winds help to mix the segmented layers of hot and cold air, thus preventing the temperature inversion. In contrary, weak winds lead to the segmentation of the air and hence temperature inversion is more likely to occur. Winds along with rains also prevent the segmentation of air.

3.10 Solar Radiation

Solar radiation is the radiant energy emitted by the sun from a nuclear fusion reaction that creates electromagnetic energy. The spectrum of solar radiation is close to that of a black body with a temperature of about 5800 K. About half of the radiation is in the visible short-wave part of the electromagnetic spectrum. The other half is mostly in the near-infrared part, with some in the ultraviolet part of the spectrum. The units of measure are Watts per square meter (W/m^2). A *solar radiation* sensor measures solar energy from the sun.

3.10.1 Nature and Properties of Solar Radiation

Solar radiation is the radiant energy emitted by the sun, particularly electromagnetic energy. About half of the radiation is in the visible short-wave part of the electromagnetic spectrum. The other half is mostly in the near-infrared part, with some in the ultraviolet part of the spectrum.

3.10.2 Components and its Impact on Plant Growth

- **Direct radiation:** Direct radiation is defined as radiation that has not been scattered in the atmosphere and hence is directionally fixed, emanating from the disc of the Sun. Direct radiation is also sometimes called "beam radiation" or "direct beam radiation," since it comes from the disc of the sun. It is used to describe solar radiation that travels in a straight line from the sun to the surface of the earth.
- **Diffuse radiation:** It is also termed as scattered radiation. It is the radiation that has experienced scattering processes in the atmosphere. It describes sunlight that has been scattered by molecules and particles in the atmosphere but has still

reached the surface of the earth. Direct radiations have a definite straight direction; however, diffuse radiation takes up any of the random path. When the radiation is direct, the rays travel in the same direction; an object can block them all at once. This is why *shadows* are only produced when *direct radiation* is blocked.

3.10.3 Ratio of Direct to Diffuse Radiation

When the sky is clear and the sun is very high in the sky, i.e., around the noon, direct radiation is around 85% of the total insolation striking the ground and diffuse radiation is about 15%. As the sun lowers in the sky, the percentage of diffused radiation increases until it reaches 40% when the sun is 10° above the horizon. Atmospheric conditions such as clouds and pollution also increase the percentage of diffuse radiation. On an extremely overcast day, almost 100% of the solar radiation is diffuse radiation. The greater the value of diffuse radiation, the lower will be the percentage of total insolation.

3.10.4 Factors Affecting Direct/Diffuse Ratio

There are primarily two main factors that affect the insolation at a particular place, i.e., altitude and climate.

Altitude: The percentage of diffuse radiation is much higher at lower altitudes than at higher altitudes, owing to the fact that at higher altitudes, direct radiation suffers less scattering due to fewer molecules of air, and pollution is very less or negligible. However, in case of low-lying areas or the places at lower altitudes, radiation is scattered more due to gas molecules and pollutants, hence the percentage of diffuse radiation is higher.

Climate: During sunny days, the percentage of direct radiation is more in comparison to the cloudy days, except for the mornings and the evenings when the diffuse radiation overpowers the direct beam radiation. Since in cloudy days the direct transmission of direct radiation is blocked by the cloud cover, the value of diffuse radiation is increased. The scattering percentage of clouds depends mainly on the height and the type of the cloud. In addition, during winter the percentage of diffuse radiation on sunny days is much higher than the sunny days of summer, which is mainly attributed to the fact that in short days of winter the sun's rays are at much slanted angle and therefore they undergo more scattering process than the long days of summer where sun's radiations are straight and vertical. The sunniest places tend to have less seasonal variation in the ratio of diffuse and direct radiation.

3.10.5 Reflected Radiation

Reflected radiation constitutes that part of solar radiation that is reflected off non-atmospheric objects such as the ground. Asphalt reflects about 4% of the light that strikes it, whereas a lawn reflects about 25%. However, solar panels tend to be tilted away from the direction of the reflected light, and reflected radiation rarely accounts for a significant part of the sunlight striking their surface. Figure 3.10 shows the absorbed scattered and reflected radiations that fall on the surface of the earth.An exception is extremely snowy conditions, which can sometimes raise the percentage of reflected radiation quite high. Fresh snow reflects 80–90% of the radiation striking it. In Fairbanks, Alaska, USA (64.5° North), there is still snow on the ground in April and May and the reflected radiation portion of the total radiation can be 25%.

3.11 Solar Radiation and Crop Production

Solar radiation is radiant (electromagnetic) energy from the sun. It provides light and heat for the earth and energy for photosynthesis. This radiant energy is necessary for the metabolism of the environment and its inhabitants. The three relevant bands, or ranges, along the solar radiation spectrum are ultraviolet, visible (PAR), and infrared. Of the light that reaches the earth's surface, infrared radiation makes up 49.4%,

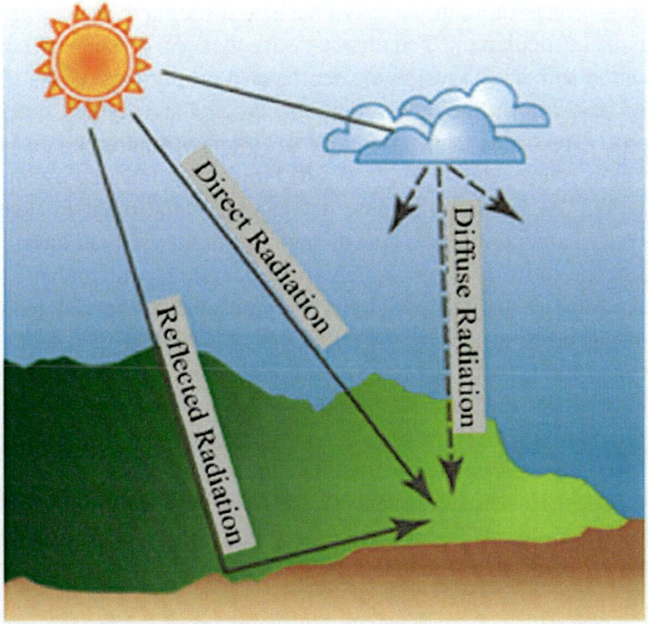

Fig. 3.10 Solar radiation and reflected radiation

while visible light provides 42.3%. Ultraviolet radiation makes up just over 8% of the total solar radiation.

Crop production is mere the exploitation of solar radiation. Generally, sunlight is divided into three major components:

1. visible light, with wavelengths between 0.4 and 0.8 μm,
2. ultraviolet light, with wavelengths shorter than 0.4 μm, and
3. infrared radiation, with wavelengths longer than 0.8 μm.

The visible portion constitutes nearly half of the total radiation received at the surface of the earth. Although ultraviolet light constitutes only a very small proportion of the total radiation, this component is extremely important.

The main advantage of infrared radiation is its heat-producing quality. Close to half of total solar radiation received at the surface of the earth is infrared.

1. **Ultraviolet:** It consists of radiations having wavelength less than 0.4 μm. It is further divided into three types, i.e., UV-A, UV-B, and UV-C.
   Ultraviolet is tolerable to the plants, and under normal condition, it is seen to increase the plant photosynthesis and release of secondary metabolites. Under high photosynthetic active radiation, UV-B light increases the net plant photosynthesis in several plant species. In both young and old plants, high PAR and UV-B can induce higher production of flavonoids. UV-A radiation also has a positive effect on the photosynthesis when plants are exposed to UV-B. However, it has been suggested that the increase in ground-level UV-B may have major deleterious effects on crop photosynthesis and productivity as a result of stratospheric ozone depletion. The direct consequences of such effects have been projected by some as a worldwide decrease in crop yields of 20–25%.
2. **Infrared radiation:** Infrared radiations are the necessary source of thermal radiations. It provides thermal energy to the plant microclimate in presence of water vapors. According to Texas A&M University, infrared light plays a part in the blooming of flowering plants. It encourages the plant to flower in indoor conditions when exposed to a specific amount of infrared radiation. According to Planta, increased infrared waves can affect the rate at which plant stems grow. A short exposure to far infrared light increases the internodal length when the exposure occurred at the end of an 8-h light period. If the plant is exposed to ordinary red light, the effect is reversed. However, a combination of far-red and red light produces the longest internodes.
   Infrared radiations can be beneficial only up to certain level, and hence too much of infrared light, especially in the far-red end of the spectrum, may actually damage plants. Heat may discolor or destroy plants, especially if they have not recently been watered. Too much infrared light may also cause plants to experience early growth spurts that reduce their health, or encourage them to flower too soon.
3. **Visible spectrum:** This spectrum lies between UV and IR and it is also termed as light. Visible light is essential for normal growth and development of plants.

Visible light is divided into violet (380–430 nm), blue (430–500 nm), green (500–570 nm), yellow (570–590 nm), orange (590–630 nm), and red (630–770 nm). Plants photosynthesize in the range between 400 and 700 nm is known as photosynthetic active radiation (PAR). However, in addition to providing energy, light also plays a key role in many other plant processes such as photomorphogenesis and photoperiodism. A wide range of signals and information for morphogenesis and many other physiological processes are triggered by light. Plant responses to light differ based on the lighting environment, season, genotype, cultivation practices, and many others. While poor light leads to plant abnormalities, high light intensity leads to abiotic stress for the plant. Light affects the production of tillers, the stability, strength, and length of culms. It affects the yield, total weight of plant structures, size of the leaves, and root development.

The quantity, quality, intensity, and duration of light are all important factors for plant growth.

- **Quality:** Light quality refers to the color or wavelength. Visible light is divided into violet (380–430 nm), blue (430–500 nm), green (500–570 nm), yellow (570–590 nm), orange (590–630 nm), and red (630–770 nm). The range in which photosynthesis of plants occurs is referred to as photosynthetic active radiation (PAR), which is between 400 and 700 nm. The regulations of photosynthesis by light quality include regulations of stomatal movement, leaf growth, chloroplast structure, photosynthetic pigment, D1 protein and its gene and photosynthetic carbon assimilation by visible light, and effect of ultraviolet light on photosystem II in plant. Chlorophyll, the green pigment in leaves responsible for absorbing the PAR, has two peaks of absorption: blue light and red light. Leaves absorb a small amount of green and reflect it back; this is why we see the color of the leaves as green only, with the exception of colored foliage.

  The role of different lights of visible spectrum is mentioned as follows:

  Blue light: It parallels to one of the absorption peaks, and hence the photosynthetic process is more efficient in presence of blue light. Blue light is responsible for vegetative and leaf growth and is important for seedlings and young plants because it helps reduce plant stretching.

  Red light: This corresponds to other peak of light absorption by the leaves. Phytochrome (a photoreceptor) within the leaves is more sensitive to and responds to red light. This light is important in the regulation of flowering and fruiting. It also helps to increase stem diameter and promotes branching.

  Blue light and red light both can promote the opening of stomata, while green light can close stomata. Blue light can improve the development of chloroplast, complex light of red, blue, and green lights can expand leaf area, and red light can increase the accumulation of photosynthesis production. Effects of different light quality differ in various plants, organs, and tissues. Blue light and far-red light can promote the accumulation of psbA gene transcription.

Different light sources distribute light differently and hence serve different purposes. Some of the artificial sources of lightning that are used in protected cultivation and indoors are as follows:

- ***Incandescent lamp***: It generates most of its light in red spectrum and only a small amount comes from blue spectrum.
- ***Fluorescent lamps***: These generate light in all the three spectrums, i.e., blue, green, and red. However, the highest level is in blue spectrum.
- ***High-pressure sodium lamp***: The highest peak is in green spectrum and is closely followed by red.
- ***Metal halide lamp***: The highest peak is in the green spectrum; the red spectrum has about half the energy peak as green and least in blue.
- ***Light emitting diodes*** *(LED)*: This type of light emits a specific wavelength of customer's choice. The manufacturer can produce these diodes in a specific color/ wavelength (monochromatic) that a customer requires.

3.12 Light Duration or Photoperiod

Photoperiodism is the response to changes in daylength that enables *plants* to adapt to seasonal changes in their environment. The number of hours of day light per day directly impacts flowering. Plants can be divided into three categories based on the required day length needed to initiate flowering:

- **Short-day plants**: These plants flower only when the day length is shorter than the night. They bloom in early spring or fall. When the day length exceeds a critical time, plants stop flowering and go into vegetative growth. Examples include chrysanthemum, cosmos, kalanchoe, poinsettia, and zinnia.
- **Long-day plants**: These plants flower when the day length is longer than the night. They flower in late spring to early summer. When the day length is shorter than a critical time, plants stop flowering and go into vegetative growth. Examples include carnation, dill, foxglove, petunia, and snapdragon.
- **Day-neutral plants**: These plants flower regardless of the day length. Instead, they usually flower after reaching certain stages of development. Examples include celery, cucumber, hydrangea, pansy, pepper, and tomato.

The rate of photosynthesis is impacted mostly by the light intensity and quality. The faster the photosynthetic rate, the faster the plant grows. The length of the day is important to know when it comes to flowering, as it directly impacts the timing of flowering for many ornamental crops. For example, it is impossible to grow a short-day plant like poinsettia to bloom in the summer under natural conditions.

3.12.1 Intensity of Light

Light intensity is usually defined as the energy hitting an area over some time period. Plants use light as a source of energy for photosynthesis. The term photosynthesis refers to the reaction between carbon dioxide and water in the presence of light to produce carbohydrates and oxygen. The rate of this process is highly dependent on the light intensity. More light intensity brings a large number of photons and hence increasing the rate of photosynthesis, but only up to a certain limit. The intensity of light can only increase the rate of photosynthesis when the other factors are not limiting. Sometimes high light intensity can be a problem in water scarce areas where water is a limiting factor. Another limiting factor that does not permit the rate of photosynthesis to go beyond the certain threshold value is the amount of chlorophyll molecules. At high light intensity, the rate of photosynthesis drops quickly as the light starts to damage the plant.

Light compensation point: It corresponds to the point when the rate of photosynthesis is equal to the rate of respiration. This means that the carbon dioxide released from respiration is equivalent to that which is taken up during photosynthesis. As the light intensity increases, the compensation point is reached. If the light intensity is increased beyond the compensation point, the rate of photosynthesis increases proportionally until the point of *light saturation* is touched, beyond which the rate of photosynthesis is no longer affected by light intensity.

Light saturation point: More light generally corresponds to higher levels of photosynthesis. However, as the light intensity increases, the photosynthetic rate eventually reaches threshold point, beyond which increase in light intensity does not increase the photosynthetic rate as seen in Fig. 3.11. This point where the light intensity does not increase the photosynthesis rate is called the light saturation point. When this point is reached, the photosynthesis rate curve becomes flat.

3.13 Solar Constant

The *solar constant* (G_{SC}) is a flux density measuring mean solar electromagnetic radiation (solar irradiance) per unit area. It is measured on a surface perpendicular to the rays, 1 astronomical unit (AU) from the Sun (roughly the distance from the Sun to the Earth).

The solar constant includes all types of solar radiation and not just the visible light (though by convention, neutrinos, being electrically neutral, do not radiate). It is measured by satellite as being 1.361 kilowatts per square meter (kW/m^2) at solar minimum (the time in the 11-year solar cycle when the number of sunspots is minimal) and approximately 0.1% greater (roughly 1.362 kW/m^2) at solar maximum.

The energy received on a unit area at the outer most boundary of the earth (atmosphere) surface held perpendicular to the sun's direction, at the mean distance between the sun and the earth. The solar constant is not a true constant. It fluctuates by as much as ±3.5% about its mean value depending on the distance of the earth

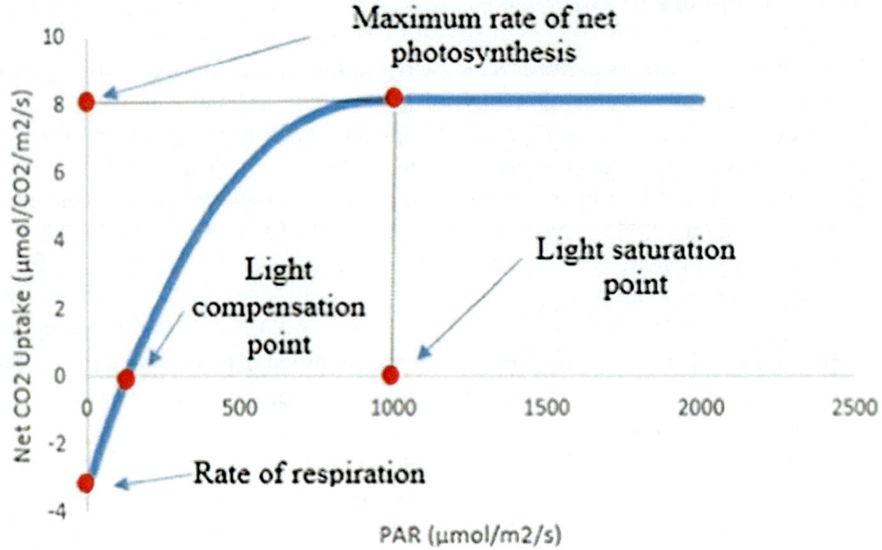

Fig. 3.11 light saturation point

from the sun. The value is 2 cal/cm^2/min (1.92 and 2.02). Recent measurements indicate a value of 1.94 cal/cm^2/min (133 wm^{-2}) [1 Langley = 1 cal]; 35% of the energy is contributed by UV and visible parts and 65% by infrared.

3.14 Depletion of Solar Radiation

The solar radiation upon entering the earth's atmosphere undergoes a number of phenomena. This phenomena deplete or filter the amount of solar radiation before reaching the surface of the earth. This filtration of solar radiation is essential for the earth to sustain life, otherwise the earth would be too hot to live in. The solar radiation upon entering the earth's atmosphere is depleted in four ways, i.e., dispersion, scattering, reflection, and absorption.

3.14.1 Dispersion

Dispersion of insolation in the atmosphere is caused by the rotation of the earth. Dispersion of insolation takes place with seasons in all latitudes; however, it is more in the latitudes of polar areas. The earth's axis is inclined at an angle of 23.5°. This inclination causes the sun's rays to be received on the surface of the earth at varying angles of incidence, depending on the position of the earth. When the sun's rays are not perpendicular to the surface of the earth, the energy becomes more dispersed or spread out over a greater surface area as seen in Fig. 3.12.

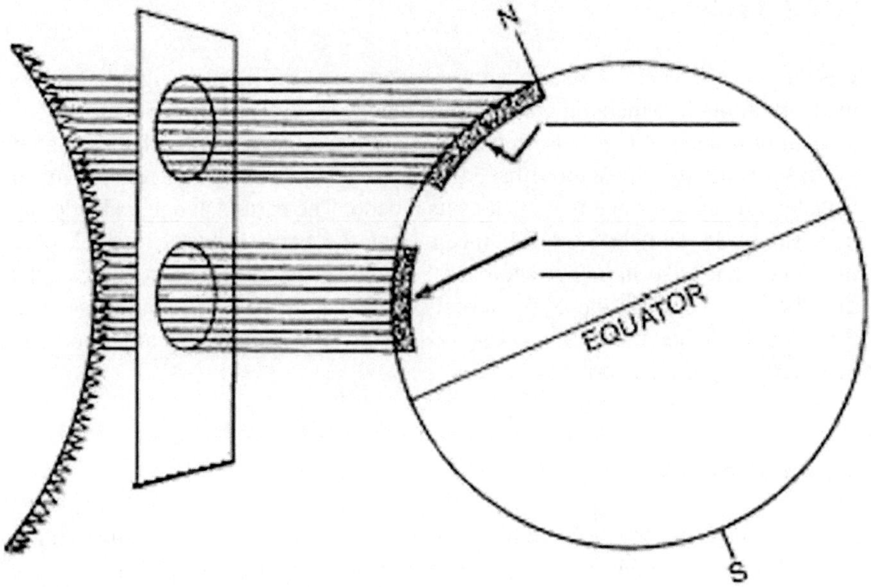

Fig. 3.12 Dispersion of solar radiation

If the energy reaching the atmosphere is constant and is dispersed over a greater area, the amount of energy at any given point within the area decreases, and considerably the temperature decreases.

3.14.2 Scattering

About 25% of the incoming solar radiation is scattered or diffused by the atmosphere. Scattering is a phenomenon that occurs when solar radiation passes through the air and some of the wavelengths are deflected in all directions by molecules of gases, suspended particles, and water vapors. These suspended particles then act like a prism and produce a variety of colors. Various wavelengths and particles sizes result in complex scattering effects that produce the blue sky. Scattering is also responsible for the red sun and varying cloud colors at sunset and sunrise, and many other optical phenomena. However, scattering does not always produce dramatic settings. Sometimes under certain radiation wavelength and particle size conditions, all that can be seen are white clouds and a whitish haze. This occurs when there is high moisture content (large particle size) in the air and this is called diffuse reflection. Diffuse sky radiation accounts about two-thirds of the total scattered radiation. However, it is almost 100% for polar stations during winter.

3.14.3 Reflection

It is the process whereby a surface turns a portion of the incident light back into the medium through which the radiation came from. Every surface reflects certain amount of insolation that falls on its surface into the space. Earth reflects an average of 36% of the total insolation that falls on it. The percentage of reflectivity of all wavelengths on a surface is known as its albedo. The earth's average albedo ranges in between 36 and 43%, which means that the earth reflects 36–43% of total insolation that falls on its surface back into the space. However, while calculating the albedo of the earth, the assumption is made that the average cloudiness over the earth is 52%. When the earth is considered as a whole, clouds play an important role in determining the albedo value.

3.14.4 Absorption

In the troposphere, the absorption of solar radiation occurs in the visible and near-infrared regions, owing primarily to H_2O, CO_2, O_2, and O_3. The absorption in the visible spectrum, however, is relatively weak. The depletion of solar radiation is dominated by ozone absorption in the UV. Aerosols scatter and absorb solar radiation. Carbonaceous aerosols (black carbon and organics) absorb and scatter solar radiation. The presence of black carbon aerosols results in the absorption of solar radiation, which reduces the solar radiation reaching the surface of the earth. At the same time, these aerosols absorb the upward solar radiation reflected from below and reduce the solar radiation reflected to space.

3.15 Albedo

It is the percentage of light that is reflected by a body or surface. It is commonly used in astronomy to describe the reflective properties of planets, satellites, and asteroids. However, its important application can be found in climatology to predict the climate change trends. High albedo designates that much of the incident solar radiation is reflected rather than absorbed and vice versa. For example, albedo decreases in the Arctic have increased heat absorption at the surface.

Albedo is a dimensionless feature and its value varies from 0 in perfectly black body to 1 in perfectly white body. The darker a body, the more it will absorb the radiations falling on it, and the lower will be the albedo number. Similarly, the white surface reflects most of the radiations falling on it, and hence the albedo value will be more, i.e., near to 1. Oceans, lakes, and forests reflect relatively small fractions of the incident sunlight and have low albedos. Snow, sea ice, and deserts reflect relatively large fractions of the incident sunlight and have large albedos.

Albedo is usually classified into two general types: normal albedo and bond albedo.

3.15.1 Normal Albedo

Normal albedo refers to the normal reflectance of a body. When the light falls on the surface of the body, it reflects certain amount of that radiation back to the space. Hence, it is a measure of a surface's relative brightness when irradiated and observed vertically, e.g., the normal albedo of snow is nearly 1.0, whereas that of charcoal is about 0.04.

3.15.2 Bond Albedo

It is named after an American astronomer, George P. Bond, who published a comparison of the brightness of the Sun, the Moon, and Jupiter in 1861. Bond albedo refers to the percentage of total incident solar radiation reflected by the planet back to space. It is a measure of the planet's energy balance. The value of Bond albedo depends on the spectrum of the incident radiation because such albedo is defined over the entire range of wavelengths, and the albedo values of different entities are mentioned in Table 3.1. Earth-orbiting satellites have been used to measure the earth's Bond albedo and the most recent value obtained is approximately 0.29. The Moon, which has a very weak atmosphere and no clouds, has an albedo of 0.12, whereas Venus, which is covered by dense clouds, has an albedo of 0.75.

The factors on which albedo depends are given as follows:

- **Angle of incidence:** Albedo is not an intrinsic property of a surface. Instead, for any surface, the albedo depends on the spectral and angular distributions of the incident light, which in turn are governed by atmospheric composition and the direction of the beam of light from the sun. When the sun is at the angles of 40° and more, the albedo of the water is fairly constant. However, as the angle decreases below 40°, the albedo percentage increases significantly. It is observed that at the angle of 10° the albedo value is 0.5 and at the angle of 0° it reaches to

Table 3.1 The approximated range of natural surfaces

Natural surface types	Approximate albedo value
Blackbody	0
Forest	0.05–0.2
Grassland and cropland	0.1–0.25
Dark-colored soil surface	0.1–0.2
Sand	0.2–0.4
Mean albedo of the earth	0.36
Light-colored soil surface	0.4–0.5
Dry salt cover	0.5
Fresh, deep snow	0.9
Water	0.1–1
Absolute white surface	1

1.0, e.g., the glare coming off the water in early morning or in the evening before sunset.

- **Texture and particle size:** Surface roughness defines the type of reflection. Shiny, smooth surfaces such as water body, plant leaves, or wet soil surfaces may be near-perfect, specular reflectors, which may reflect well and show relatively high albedo for lower sun angles. Rough surfaces represent lower albedo values, especially when sun angle is low and the shading effect lowers the reflection. There are measurable differences in the surface roughness among soil textural classes. Fine textured, dry soils with small particle size produce high albedo due to relatively smooth surface. However, clayey soils are often wet, and soil moisture absorbs the incident radiation and decreases albedo. Conversely, dry, coarse-textured soils with relatively large particles (sand grains) reflect larger portions of the incident radiation than clayey soils.
- **Soil moisture content**: Albedo varies greatly with soil moisture content. The higher the soil moisture content, the darker the color and the lower the albedo. However, this relationship is valid only for soil moisture contents up to the field capacity. Beyond field capacity, the increase in soil moisture content does not darken the color any more, but starts building up a water sheet on the aggregate surface, creating a shiny and better reflecting surface that increases the reflectance and thus the albedo. This phenomenon is the major reason for differences in the albedo among soils of different textural classes as depicted in Fig. 3.13. Clayey soils can maintain high moisture content in the presence of water supply, while sandy-textured soils drain and dry out much more rapidly. Due to the differences in the resulting soil moisture content between the texture classes, there are differences in the reflectance and absorbance characteristics and so in the albedo.
- **Organic matter content:** Soil organic matter content increases the absorbance of the soil. Organic matter makes the soil darker, decreasing the albedo level. Thus, the higher the organic matter content, the lower the albedo.
- **Soluble salts:** Salt crust on the surface increases the albedo dramatically. As a result, mapping of salt-affected area with remotely sensed images is a very powerful tool for soil surveyors.
- **Type and condition of vegetation:** The type and the condition of the vegetation have a strong impact on the surface albedo. Forest vegetation with multilevel canopy has a low albedo because the incident radiation can penetrate deeply into the forest canopy where it bounces back and forth between the branches and leaves and get trapped by the canopy. The albedo for grassland and cropland ranges between 0.1 and 0.25.
- **Season:** Contemporary agricultural practices give rise to large variations in albedo from season to season as the land passes through the cycle of tilling, planting, crop growth, and harvest.
- **Iron oxides:** Iron oxides increase reflectivity in the red portion of the spectrum while decreasing it in the blue–green and infrared portions.

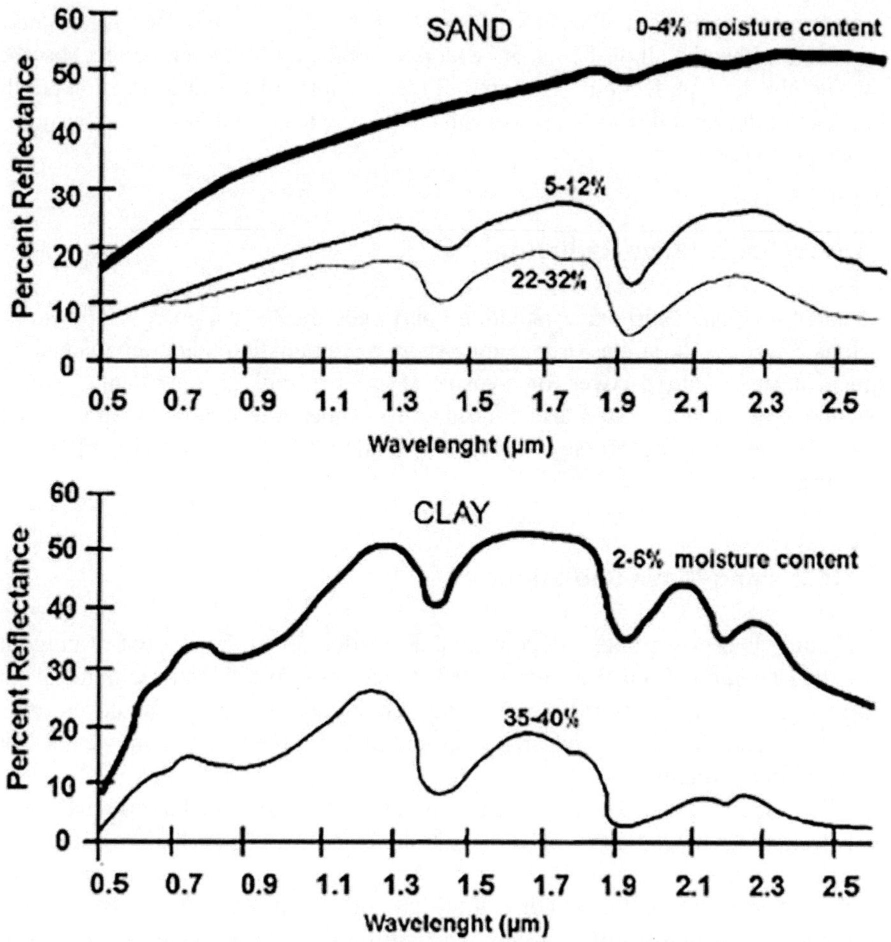

Fig. 3.13 Albedo among soils of different textural classes

3.16 Thermal Radiation

Thermal radiation is the process by which energy in the form of electromagnetic radiation is emitted by the heated surface in all directions and travel directly to the point of absorption at the speed of light. Thermal radiation does not require any intervening medium to carry. It ranges in wavelength from the longest infrared rays through the visible-light spectrum to the shortest ultraviolet rays. The intensity and distribution of radiant energy within this range is governed by the temperature of the emitting surface.

The rate at which a body radiates (or absorbs) thermal radiation depends on the nature of the surface as well. Objects that are good emitters are also good absorbers. A blackened surface is an excellent emitter as well as an excellent absorber. If the same surface is silvered, it becomes a poor emitter and a poor absorber. A blackbody is one that absorbs all of the radiant energy that falls on it. Such a perfect absorber would also be a perfect emitter. Some of the examples of thermal radiation are the heating of the earth by the sun's rays and the heating of a room by the combustion of coal, among others.

3.16.1 Short-Wave Radiation

Incoming (solar) short-wave radiation constitutes the largest energy input to the climate system. Variations of incoming short-wave radiation (e.g., due to changing incident angle, cloud cover, or aerosol effects) are thus an important source of temperature variability and also contribute to temperature trends. Incoming short-wave radiation is the primary source of radiation that heats up the earth and its atmosphere.

3.16.2 Long-Wave Radiation

Incoming long-wave radiation contains information about the amount of outgoing thermal radiation from the earth's surface that is absorbed and reemitted in the atmosphere. It depends on atmospheric conditions such as cloud cover, water vapor, or greenhouse gas concentrations and includes both short-term weather and long-term climate effects.

Outgoing long-wave radiation (OLR) is energy emitted from the earth in the form of electromagnetic radiation that passes through the atmosphere and into space in the form of thermal radiation. OLR is primarily sensitive to near-surface and atmospheric temperatures, the humidity of the air, and the presence of clouds, which are related to the intensity of convective activity as well as the latitude and altitude dependency of the variability.

OLR is a measure of the amount of energy emitted to space by the earth's surface, oceans, and atmosphere. As such, it is a critical component of the earth's radiation budget. In the earth's climate system, long-wave radiation involves processes of absorption, scattering, and emissions from atmospheric gases, aerosols, clouds, and the surface. The flux of energy transported by outgoing long-wave radiation is measured in W/m^2 and can be measured by a satellite. Figure 3.14 depicts the distribution of annual mean absorbed and emitted and net radiations over the surface of the earth.

At the average temperature of 12 °C, 99% of the total outgoing energy is emitted in the form of infrared. From this, 90% of the energy is reabsorbed back by the atmosphere in varied wavelengths. Water vapors absorb in the range of 5.3–7.7 μ and beyond 20 μ, ozone in the range of 9.4–9.8 μ, CO_2 in the range of 13.1–16.9 μ, and clouds absorb in all wavelengths. Long-wave radiation escapes to the space between 8.5 and 11 μ and this is known as the atmospheric window. Here the atmosphere acts as transparent medium and there is relatively little absorption of

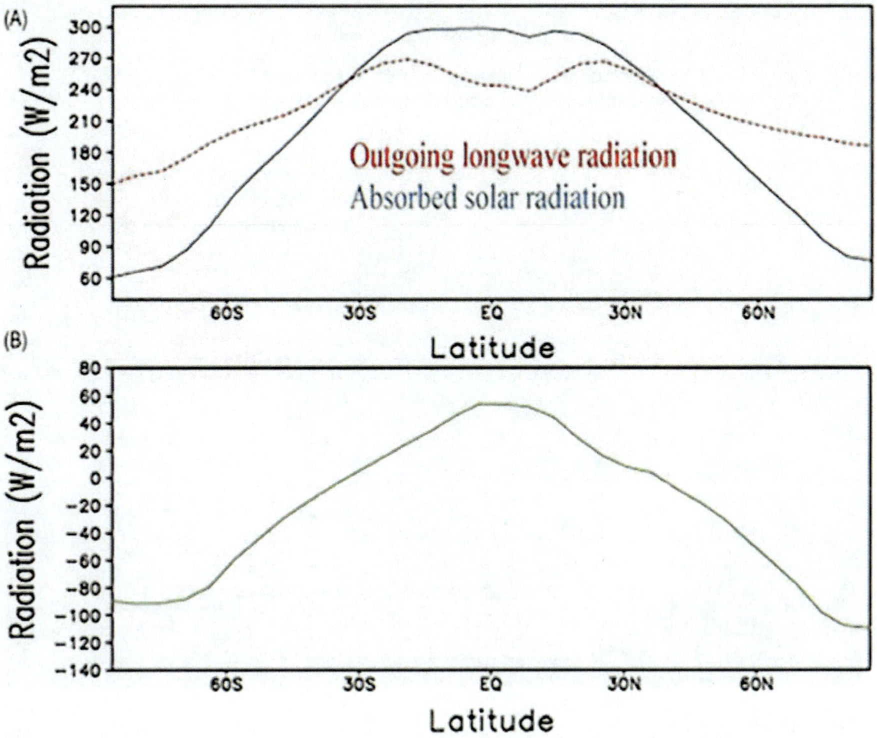

Fig. 3.14 Distribution of annual mean absorbed and emitted (**a**) and net (**b**) radiations over the earth

radiation. This spectral region finds its application in microwave remote sensing where it is used to monitor the features of the sky in the event of overcast sky. A large part of the radiation absorbed by the atmosphere is sent back to the earth's surface at night known as counter-radiation and it prevents the earth's surface from excessive cooling at night.

3.16.3 Net Radiation

The earth's net radiation, sometimes called net flux, is the balance between incoming and outgoing energy at the top of the atmosphere. It is the total energy that is available to influence the climate. Energy enters the system when sunlight penetrates the top of the atmosphere. Energy is released in two ways: reflection by clouds, aerosols, or the earth's surface; and thermal radiation where heat is emitted by the surface and the atmosphere, including clouds. The global average net radiation must be close to zero over the span of a year or else the average temperature will rise or fall. Figure 3.15 shows the schematic distribution of the radiation (energy) balance (budget) in the earth–atmosphere system.

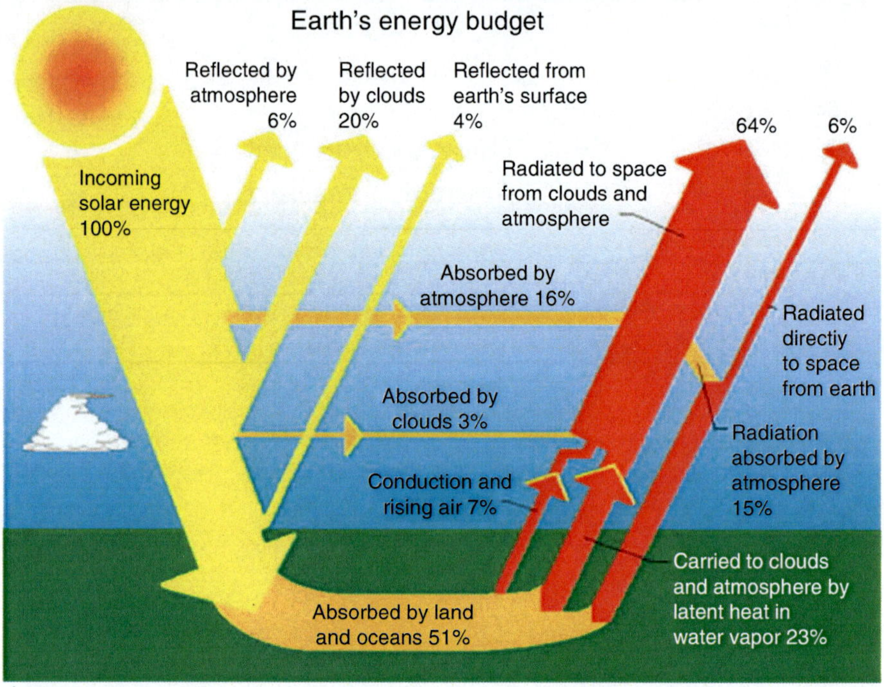

Fig. 3.15 Schematic distribution of the radiation (energy) balance (budget) in the earth–atmosphere system. The incoming (short-wave) solar radiation is shown in yellow and the outgoing (long-wave or thermal) radiation in red

Over the course of a year, the most obvious pattern is seasonal changes in net radiation. Incoming sunlight increases in the hemisphere, resulting in summer, which makes the energy imbalance strongly positive (more watts of energy coming in than going out). As the September equinox approaches, a zone of positive net radiation is nearly centered over the equator, and energy deficits lie over the poles. As the season turns to winter, net radiation becomes negative in the Northern Hemisphere and positive in the Southern Hemisphere. The pattern reverses on the March equinox.

Over the course of the year, there is a net energy surplus at the equator and a net energy deficit at the poles. This equator-versus-pole energy imbalance is the fundamental driver of atmospheric and oceanic circulation.

Bibliography

Ambient weather.com (n.d.) https://ambientweather.com/solarradiation.html

Anonymous (n.d.) https://astrolabesailing.com/2016/09/20/meteorology-what-about-the-wind/

Arrhenius S (1896) XXXI. On the influence of carbonic acid in the air upon the temperature of the ground. London, Edinburgh, Dublin Philosophical Magazine J Sci 41(251):237–276

Baylis AD, Wright ITJ (1990) The effects of lodging and a paclobutrazol–chlormequat chloride mixture on the yield and quality of oilseed rape. Ann Appl Biol 116(2):287–295

Berry PM, Spink JH (2006) A physiological analysis of oilseed rape yields: past and future. J Agric Sci 144(5):381–392

Berry PM, Sterling M, Spink JH, Baker CJ, Sylvester-Bradley R, Mooney SJ, Tams AR, Ennos AR (2004) Understanding and reducing lodging in cereals. Adv Agron 84(04):215–269

Britannica E (2019) The editors of Encyclopaedia Britannica. de la Enciclopedia Británica

Britannica, T. Editors of Encyclopaedia (2020a). Wind. Encyclopedia Britannica. https://www.britannica.com/science/wind

Britannica, T. Editors of Encyclopaedia (2020b) Land breeze. Encyclopedia Britannica. https://www.britannica.com/science/land-breeze

Britannica, T. Editors of Encyclopaedia (2021). Atmospheric pressure. Encyclopedia Britannica. https://www.britannica.com/science/atmospheric-pressure

Britannica (n.d.) https://www.britannica.com/science/westerlies

Brittanica (n.d.-a) http://cdn.britannica.com/s:690x388,c:crop/00/102500-050-2A2e7E20/circulationpatterns-Earth-surface-diagram.jpg

Brittanica (n.d.-b) https://www.britannica.com/science/mountain-breeze

Budyko MI (1969) The effect of solar radiation variations on the climate of the earth. Tellus 21(5):611–619

Burgess AJ, Retkute R, Preston SP, Jensen OE, Pound MP, Pridmore TP, Murchie EH (2016) The 4-dimensional plant: effects of wind-induced canopy movement on light fluctuations and photosynthesis. Front Plant Sci 7:1392

Carter PR, Hudelson KD (1988) Influence of simulated wind lodging on corn growth and grain yield. J Prod Agric 1(4):295–299

Chen M, Chori F (2004) Light signal transduction in higher plants. Annu Rev Genet 38:87–117

Chen J, Wang W, Fang J, Varahramyan K (2004) Variable-focusing microlens with microfluidic chip. J Micromech Microeng 14(5):675

Cleugh HA, Miller JM, Böhm M (1998) Direct mechanical effects of wind on crops. Agrofor Syst 41(1):85–112

Compo GP, Sardeshmukh PD (2009) Oceanic influences on recent continental warming. Clim Dyn 32(2):333–342

El-Halabi S, Basheer M and Al-Saleh F, (2014) Design of self-sustained vegetable farm house using integrated power production and water cooling system

ESALQ (n.d.) http://www.esalq.usp.br/lepse/imgs/conteudo_thumb/How-does-the-level-of-light-affect-the-rate-of-photosynthesis.pdf

Fiscus EL, Booker FL (1995) Is UV-B a hazard to crop photosynthesis and productivity? Results of an ozone-UV-B interaction study and model predictions. Photosynth Res 43:81–92

Fitzroy R (1863) The weather book: a manual of practical meteorology, vol 2. Longman, Green, Longman, Roberts, & Green, England

Fondriest Environmental Learning Center (2014) Solar radiation and photosynthetically active radiation. Fundamentals of Environmental Measurements

FT Exploring Science and Technology (n.d.) https://www.ftexploring.com/solar-energy/direct-and-diffuse-radiation.htm

Glickman TS, Zenk W (2000) Glossary of meteorology. AMS (American Meteorological Society), Boston

Gov.Uk (n.d.) https://www.metoffice.gov.uk/weather/learn-about/weather/types-of-weather/wind/types-of-wind

Graham E, Sarazin M, Matzler C (2011) Using re-analysis model data ("FriOWL") to analyse climate trends related to astroclimatology of Paranal and La Silla. Rev Mex Astron Astrofís (Ser Conf) 41:12–15

Harman JR (1987) Mean monthly North American anticyclone frequencies, 1950–79. Mon Weather Rev 115:2840–2848

Hunker.com (n.d.) https://www.hunker.com/12000225/the-effect-of-infrared-light-on-plant-growth

Imperiali N, Chiriboga X, Schlaeppi K, Fesselet M, Villacrés D, Jaffuel G, Bender SF, Dennert F, Blanco-Pérez R, Van der Heijden MG, Maurhofer M (2017) Combined field inoculations of Pseudomonas bacteria, arbuscular mycorrhizal fungi, and entomopathogenic nematodes and their effects on wheat performance. Front Plant Sci 8:1809

Kane RP, Buriti RA (1997) Latitude and altitude dependence of the interannual variability and trends of atmospheric temperatures. Pure Appl Geophys 149(4):775–792

Kashiwagi T, Ishimaru K (2004) Identification and functional analysis of a locus for improvement of lodging resistance in rice. Plant Physiol 134(2):676–683

Kozai T (2016) Why LED lighting for urban agriculture? In: LED lighting for urban agriculture. Springer, Singapore, pp 3–18

Lange OL, Nobel PS, Osmond CB, Ziegler H (1981) Introduction: perspectives in ecological plant physiology. In: Physiological plant ecology I. Springer, Berlin, Heidelberg, pp 1–9

Langre E, d. (2008) Effects of wind on plants. Annu Rev 40:141–168

Lohmann U, Feichter J (2005) Global indirect aerosol effects: a review. Atmos Chem Phys 5(3): 715–737

Lopez JC (2021) Influence of light on crop growth. promix: https://www.pthorticulture.com/en/training-center/influence-of-light-on-crop-growth/

National Ocean Service (n.d.) https://oceanservice.noaa.gov/education/tutorial_currents/04currents1.html

National Oceanic and Atmospheric Administration (n.d.) https://www.weather.gov/jetstream/seabreeze

Niklas KJ, Spatz HC (2012) Plant physics. In: Plant physics. University of Chicago Press, Chicago

Ouzounov D, Liu D, Chunli K, Cervone G, Kafatos M, Taylor P (2007) Outgoing long wave radiation variability from IR satellite data prior to major earthquakes. Tectonophysics 431(1–4): 211–220

Philipona R, Dürr B, Ohmura A, Ruckstuhl C (2005) Anthropogenic greenhouse forcing and strong water vapor feedback increase temperature in Europe. Geophys Res Lett 32:L19809

Pithan F, Mauritsen T (2014) Arctic amplification dominated by temperature feedbacks in contemporary climate models. Nat Geosci, 7(3):181–184

Premiertech horticulture (n.d.) https://www.pthorticulture.com/media/2100/light_crops_en.jpg?width=500&height=295.09803921568624

Rajkumara S (2008) Lodging in cereals–a review. Agric Rev 29(1):55–60

rammb.cira (n.d.) http://rammb.cira.colostate.edu/wmovl/vrl/tutorials/satmanu-eumetstat/satmanu/cms/sb/backgr.htm

Reitan CH (1974) A climatic model of solar radiation and temperature change. Quat Res 4(1):25–38

Sciencedirect.com (n.d.-a) https://www.sciencedirect.com/topics/engineering/direct-radiation

Sciencedirect.com (n.d.-b) https://www.sciencedirect.com/topics/agricultural-and-biologicalsciences/photoperiodism

SciJinks (n.d.) https://scijinks.gov/trade-winds/

Setter TL, Laureles EV, Mazaredo AM (1997) Lodging reduces yield of rice by self-shading and reductions in canopy photosynthesis. Field Crop Res 49(2–3):95–106

Smith VC, Ennos AR (2003) The effects of air flow and stem flexure on the mechanical and hydraulic properties of the stems of sunflowers *Helianthus annuus* L. J Exp Bot 54(383): 845–849

Stephens GL, Webster PJ (1981) Clouds and climate: sensitivity of simple systems. J Atmos Sci 38(2):235–247

Sterling M, Baker CJ, Berry PM, Wade A (2003) An Experimental investigation of the lodging of wheat. Agric For Meteorol 119(3–4):149–165

tpub.com (n.d.) http://meteorologytraining.tpub.com/14312/img/14312_19_1.jpg

University of California, San Diego (n.d.) http://earthguide.ucsd.edu/virtualmuseum/climatechange1/08_3.shtml

University of Illinois (n.d.) http://ww2010.atmos.uiuc.edu/(Gh)/guides/mtr/prs/hght.rxml

Weebly.com (n.d.) https://easyagriculture.weebly.com/uploads/1/0/7/3/107365775/agricultural_meteorology.pdf

Wikipedia (n.d.) https://en.wikipedia.org/wiki/Westerlies

Zhang A, Wang H, Beyene Y, Semagn K, Liu Y, Cao S, Cui Z, Ruan Y, Burgueño J, San Vicente F, Olsen M (2017) Effect of trait heritability, training population size and marker density on genomic prediction accuracy estimation in 22 bi-parental tropical maize populations. Front Plant Sci 8:1916

Zheng J, Hu MJ, Guo YP (2008) Regulation of photosynthesis by light quality and its mechanism in plants. J Appl Ecol 19(7):1619–1624

Atmospheric Humidity

4

Abstract

Atmospheric humidity is the concentration or total amount of water contained in the column of air. Vapor in the air is a determinant of weather and is an important metric used in weather forecasts. It is simply the amount of water vapor or humidity present in the air, leading to many climatic events. It acts as an indicator of the likelihood of precipitation, as well as other related meteorological events, with profound impact on crop production. This chapter, apart from atmospheric humidity, also deals with evaporation, transpiration, cloud formation, precipitation, artificial rainmaking, and other related climatic events. Wind and monsoons in context to Indian agriculture has also been discussed in detail.

Keywords

Atmospheric humidity · Vapor pressure · Dew · Fog · Mist · Frost · Cloud · Rain · Snow · Sleet · Hail · Artificial rainmaking · Wind · Monsoon

4.1 Atmospheric Humidity

Humidity is the amount of water vapor in the air. It is the concentration or total amount of water contained in the column of air. If there is a lot of water vapor in the air, the humidity will be high. The higher the humidity, the wetter it feels outside. On the weather reports, humidity is usually explained as relative humidity.

L. Ahmad et al., *Climate Change and Agrometeorology*,
https://doi.org/10.1007/978-981-99-4863-5_4

4.2 Relative Humidity

Relative humidity (RH) (expressed as a percent) measures water vapor, but it is always relative to the temperature of the air. In other words, it is a measure of the actual amount of water vapor in the air compared to the total amount of vapor that can exist in the air at its current temperature. Relative humidity is the ratio between the amount of water vapor present in a given volume of air and the amount of water vapor required for saturation under fixed temperature and pressure. In other terms it is the ratio of the air's water vapor content to its maximum water vapor capacity at a given temperature expressed in percentage. The relative humidity gives only the degree of saturation of air. The relative humidity of saturated air is 100%. At 30 °C (86 °F), 4% of the volume of the air may be occupied by water molecules, but when the air is colder than −40 °C (−40 °F), less than one-fifth of 1% of the air molecules can be water.

4.2.1 Saturation

In meteorology, saturation is the state of the atmosphere in which air comprises the maximum amount of water in the form of water vapor that it can hold at a specific temperature and air pressure. This corresponds to the relative humidity of 100%; temperature and dew point are equal, and evaporation of water stops. The moisture present in the air at saturation fluctuates with many other factors in which temperature and pressure are of primary importance. While increasing the temperature, the ability of the air to hold water vapor also increases; however, it is reverse in the case of pressure. Hence, warm air always holds more moisture than cooler air.

Saturation, therefore, refers to the condition at which air has reached its maximum capacity for holding dissolved moisture as water vapor. The saturation point is also known by a much more familiar term, the dew point, which refers to the temperature at which the air has become saturated. When the air is at a condition of saturation and the temperature falls or the atmospheric pressure rises, the air will be unable to hold the dissolved water vapor in suspension and some will be forced out as liquid water. This process is known as the formation of dew. As water molecules are forced out of suspension, they adhere to surfaces and molecular attraction causes them to collect together, forming the water droplets we know as dew.

Other terms related to humidity are as follows:

- **Absolute humidity**: The actual mass of water vapor present in a given volume of moist air.
- **Specific humidity**: Weight of water vapor per unit weight of moist air.
- **Mixing ratio**: The ratio of the mass of water vapor contained in a sample of moist air to the mass of dry air.
- **Dew point temperature**: It is the temperature to which a given parcel of air must be cooled in order to become saturated at a constant pressure and water vapor. In

this case, the invisible water vapor begins to condense into visible form such as water droplets.

- **Vapor pressure deficit**: It is the difference between the saturated vapor pressure (SVP) and actual vapor pressure (AVP) at a given temperature. This is another measure of moisture in the atmosphere, which is useful in crop growth studies. When air contains all the moisture that it can hold to its maximum limit, it is called as saturated air; otherwise, it is unsaturated air at that temperature. The vapor pressure created at this temperature under saturated conditions is known as vapor pressure or saturated vapor pressure (SVP).

4.2.2 Units of Measurement

Absolute humidity is expressed as grams of water vapor per cubic meter or cubic feet, i.e., g/kg; however, relative humidity is unitless as it is expressed in percent.

4.3 Vapor Pressure

Vapor pressure refers to the amount of pressure exerted by the molecules of given vapor. If the vapor is in pure state, then the vapor pressure is taken as total pressure exerted by the given, defined vapor onto the walls of container. If the vapors are in mixture form, i.e., combination of many gases, then the vapor pressure is taken as partial vapor pressure of each gas from the given mixture exerted onto the walls of container. In meteorology, vapor pressure is used almost exclusively to denote the partial pressure of water vapor in the atmosphere. The vapor pressure is directly related to number of water molecules present in the atmosphere at a given time. It is measured by the units of Pascal.

4.4 Importance of Humidity on Crop Plants

The humidity is not a self-governing factor. It is closely linked to rainfall, wind, and temperature. However, it is one of the primary factors that play a significant role in crop production. Relative humidity (RH) directly influences the water relations of plant and indirectly affects leaf growth, photosynthesis, pollination, occurrence of diseases, and finally economic yield. The dryness of the atmosphere as represented by saturation deficit (100-RH) reduces dry matter production through stomatal control and leaf water potential.

4.4.1 Crop Choice

The humidity determines the type of crop grown in a specific area.

4.4.2 Leaf Growth

1. Leaf growth depends not only on synthetic activities resulting from biochemical process but also on the physical process of cell enlargement.
2. Cell enlargement occurs as a result of turgor pressure developed within the cells.
3. Turgor pressure is high under RH due to less transpiration. Thus, leaf enlargement is high in humid areas.

4.4.3 Water Requirements of Plants

1. It affects the internal water potential of plants that can influence certain physiological phenomena in crop plants including transpiration. The humidity is a major determinant of potential evapotranspiration that determines the water requirement of crops.
2. High humidity reduces irrigation water requirement of crops as the evapotranspiration losses from crops affected by atmospheric humidity.
3. High humidity can prolong the survival of crops under moisture stress. However, very high or very low relative humidity is not conducive to higher yields of crops.

4.4.4 Photosynthesis

1. Photosynthesis is indirectly affected by RH. When RH is low, transpiration increases, causing water deficits in the plant.
2. Water deficits cause partial or full closure of stomata and increase mesophyll resistance, blocking entry of carbon dioxide.

4.4.5 Pollination

1. Moderately low air humidity is favorable for seed set in many crops, provided soil moisture supply is adequate, e.g., seed set in wheat was high at 60% RH compared to 80% when water availability in the soil was not limiting.
2. At high RH, pollen may not be dispersed from the anthers.

4.4.6 Pests

There are harmful effects of high humidity. It enhances the growth of some saprophytic and parasitic fungi, bacteria, and pests, the growth of which causes extensive damage to crop plants, e.g., (1) blight disease on potato, (2) the damage caused by thrips and jassids on several crops. High relative humidity also favors easy germination of fungal spores on plant leaves, e.g., apple scab.

4.4.7 Grain Yield

1. High humidity at grain filling reduces the crop yields.
2. A very high relative humidity is beneficial to maize, sorghum, sugarcane, and so on, while it is harmful to crops such as sunflower and tobacco.
3. For almost all the crops, it is always safe to have a moderate relative humidity of above 40%.

4.4.8 Very High Relative Humidity

- Reduces evapotranspiration;
- Increases heat load of plants;
- Stomatal closure;
- Reduced CO_2 uptake;
- Reduced transpiration influences translocation of food materials and nutrients;
- Moderately high RH of 60–70% is beneficial;
- Low RH increases the evapotranspiration.

4.4.9 Variation in Humidity

1. Absolute humidity is highest at the equator and minimum at the poles.
2. Absolute humidity is minimum at sunrise and maximum in afternoon from 2 to 3 pm and the diurnal variations are small in desert regions.
3. The relative humidity is maximum at about the sunrise and minimum between 2 and 3 pm.
4. The behavior of relative humidity differs a lot from absolute humidity. At the equator it is at a maximum of 80% and around 85% at the poles. However, near horse latitudes it is around 70%.

4.5 Evaporation, Transpiration, and Condensation

4.5.1 Evaporation

Evaporation is a process by which water in liquid form changes to gaseous form. It may be defined as a water loss in vapor form from soil/water surface. The energy used in this process is called latent energy. The energy of 540 cal/g of water is used in this process. Evaporation is measured by an instrument called evaporimeter. Evaporation is most important way for plants to lose heat via stomata and plays important role in evacuating heat. Evaporation depends on humidity, temperature, liquid exposed surface, and pressure of wind.

Water is first absorbed through the very tip of the root. Next, water moves either between the cell membranes or directly through cells of the plant. Eventually, it reaches an area called endodermis, as stated by Sterling. The endodermis works similar to a paper towel soaking the water upward. Once it crosses this area, it is in what is commonly known as the stem of the plant. This area moves water rapidly toward the plant's leaves. At this point, stomata react to environmental cues and release water to the surface of the leaf where it evaporates. Plants use, on average, approximately 5% of the light for photosynthesis. About 60% of the light that reaches the plant is converted into heat and should be evacuated by the plant mainly through evaporation. Heating up water from 0 to 100 °C (= 0.418×10^6 J/kg) does not cost as much energy as letting water evaporates (2.26×10^6 J/kg). Therefore, water evaporation by a plant is a very efficient way of evacuating heat. The condition, however, is that the stomata must stay open; otherwise, it will be difficult for the plant to evacuate its heat. In this case, the plant temperature can increase too much, which may result in plant damage. An example of this is chlorophyll damage, as a result of which the leaf can turn yellow. The plant will substitute the chlorophyll where possible, which will cost for assimilates. If the plant temperatures increase further, parts of the plant can also be irreversibly damaged. This is visible in the burned parts of the leaf. In order to keep the stomata open, it is of course important to keep the humidity in the substrate at the right level and not to let the electrical conductivity (EC) increase too much when the demand for water is high. In addition, by keeping a higher humidity percentage in the greenhouse, especially at more irradiation, too much humidity loss from the plant can be avoided. This can be achieved where possible by keeping humidity around the plant and, when the humidity diminishes, by working with air humidification. Stomata can also be kept open, where necessary, by limiting irradiation and/or by cooling the plants. Keeping the stomata open for photosynthesis is extremely important. When the stomata are completely closed, no photosynthesis will be possible. When this process halts, the plant will not grow. By keeping the humidity in the greenhouse and around the plant at the right level, the plant will be able to keep its stomata open. Then, CO_2 can be absorbed and the plant temperature can be regulated through evaporation. So, evaporation causes plant stress and soil problems. Evaporation causes upper levels of soil hard and dry, and when rainfall occurs, the water runs off instead of soaking into the ground, making plants difficult to grow. In addition, as water is a costly resource in many areas, one effect of evaporation is the cost of growing agriculture crops. Farmers not only have to pay to acquire water to irrigate crops, but they must also have to purchase expensive equipment.

4.5.2 Transpiration

Terrestrial plants transpire essentially through their whole surface. Although some parts are covered by layers (e.g., cork layers of the bark), which resist diffusion of water vapor, the greatest part of their surface must ensure an adequate gas exchange with the atmosphere. Then, through the same way carbon dioxide and oxygen diffuse into the plant, water vapor also diffuses out along the gradient into the environment. Whether transpiration is beneficial or harmful to the development of plants is a much debated question. Curtis contends that it is almost entirely harmful with essentially no helpful contribution to the physiology of plants. Miller appears to agree with Curtis when he states that "it must be admitted that its harmful effects outweigh any of the benefits that may be attributed to the process." MaxiMov seems to feel that transpiration is less helpful than it is harmful, suggesting that plants could adjust themselves to conditions that would prevail if transpiration were non-existent. Barnes in 1902 wrote that as plants developed on land, they had to contend with this necessary evil. Others have reasoned that transpiration must be beneficial since all processes of life are of value. Shull and others have shown the effect transpiration has on lowering temperatures.

Transpiration impacts speed of growth and stature of plants. Whatever the process may be by which soil solutes enter the living cells of plants, it is probably the same whether the plant is transpiring rapidly, slowly, or not at all. However, as soon as these solutes entered the xylem, they are in the transpiration stream and move upward with the water. Shull has pointed out that if molecules moved only by their own kinetic forces, the movement of trees would be extremely low. He further points out: "It is the transpiration that gives rise to the mass movement of the water in the tracheal system and in this current the ions and molecules are swept along." The advantage of transpiration here then is that it allows for a rapid distribution of mineral solutes once these have entered the plant. Furthermore, since this is true, a more rapid growth mechanism results in a greater extension of the root system with its resulting greater absorption. Hence, the rapid growth of plants is enabled by transpiration by:

- hastening salt distribution in the plant, and
- cooling the leaves and thereby aids the solubility of carbon dioxide in plants, thus increasing the rate of photosynthesis.

4.5.3 Condensation

The transformation of water vapor into water is called condensation. Condensation is caused by the loss of heat (latent heat of condensation). When moist air is cooled, it may reach a level when its capacity to hold water vapor ceases (saturation point = 100%). Then, the excess water vapor condenses into liquid form. If it directly condenses into solid form, it is known as sublimation. Figure 4.1 shows the process of sublimation.

In free air, condensation results from cooling around very small particles termed as hygroscopic condensation nuclei. Particles of dust, smoke, pollen, and salt from the ocean are particularly good nuclei because they absorb water.

Condensation also takes place when the moist air comes in contact with some colder object and it may also take place when the temperature is close to the dew point. Condensation, therefore, depends on the amount of cooling and the relative humidity of the air. After condensation, the water vapor or the moisture in the atmosphere takes one of the following forms—dew, frost, fog, and clouds.

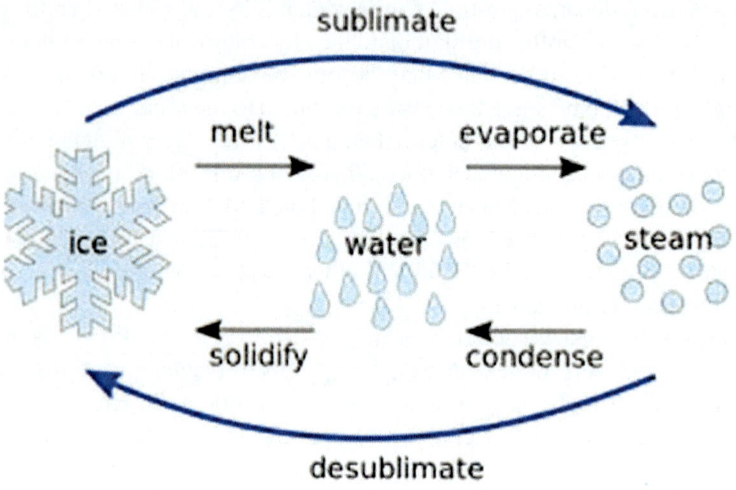

Fig. 4.1 Process of sublimation

4.5.3.1 Process of Condensation

Like all matter, water consists of molecules. In a vapor form, the molecules are energetic, fast moving, and far apart. As the vapor encounters cooler temperatures, the molecules become slower, less energetic, and closer together. When they reach a threshold energy level, the vapor changes to liquid.

4.5.3.2 Forms of Condensation

The forms of condensation can be classified on the basis of temperature at which the dew point is reached. Condensation can take place when the dew point is

- lower than the freezing point or
- higher than the freezing point.

White frost, snow, and some clouds (cirrus clouds) are produced when the temperature is lower than the freezing point. Dew, fog, and clouds result even when the temperature is higher than the freezing point. Forms of condensation may also be classified on the basis of their location, i.e.,

- at or near the earth's surface,
- in free air, and
- dew, white frost, fog, and mist come in the first category, whereas clouds are in the second category.

Dew

Deposition of moisture in the form of water droplets on cooler surfaces of solid objects, such as stones, grass blades, and plant leaves, is known as dew. The ideal conditions for its formation are clear sky, calm air, high relative humidity, and cold and long nights. For the formation of dew, it is necessary that the dew point is above the freezing point.

White Frost

Frost forms on cold surfaces when condensation takes place below freezing point (0 °C), i.e., the dew point is at or below the freezing point. The excess moisture is deposited in the form of minute ice crystals instead of water droplets. The ideal conditions for the formation of white frost are the same as those for the formation of dew, except that the air temperature must be at or below the freezing point.

Fog

When the temperature of an air mass containing a large quantity of water vapor falls all of a sudden, condensation takes place within itself on fine dust particles. So, the fog is a cloud with its base at or very near to the ground. Because of the fog and mist, the visibility becomes poor to zero. In urban and industrial centers, smoke provides plenty of nuclei that help the formation of fog and mist. Such a condition, when fog is mixed with smoke, is described as smog.

Mist

The difference between mist and fog is that mist contains more moisture than fog. In mist, each nuclei contains a thicker layer of moisture. Mists are frequent over mountains as the rising warm air up the slopes meet a cold surface. Mist is also formed by water droplets, but with less merging or coalescing. This means mist is less dense and quicker to dissipate. Fogs are drier than mist and they are prevalent where warm currents of air come in contact with cold currents. In mist, visibility is more than 1 km but less than 2 km.

Haze

Haze is traditionally an atmospheric phenomenon where dust, smoke, and other dry particles obscure the clarity of the sky. (There is no condensation in haze. Smog is similar to haze but there is condensation in smog) Sources for haze particles include farming (ploughing in dry weather), traffic, industry, and wildfires.

Smog

Smog = smoke + fog (smoky fog) is formed by the burning of large amounts of coal, vehicular emission, and industrial fumes (primary pollutants).

Clouds

Cloud is a mass of minute water droplets formed by the condensation of the water vapor in free air at considerable elevations. Clouds are formed mainly by the adiabatic cooling of air below its dew point. As the clouds are formed at some height over the surface of the earth, they take various shapes. According to their height, expanse, density, and transparency or opaqueness, clouds are classified into four types: (1) cirrus, (2) cumulus, (3) stratus, and (4) nimbus.

4.6 Clouds and Their Formation

Clouds are a set of visible ice crystals or water droplets suspended in air at higher altitudes. Cloud formation results from the presence of several contributing factors in the atmosphere, which as follows:

- The first prerequisite is water substance. Especially from a meteorological point of view, water vapor is the most important substance found in air.
- The second prerequisite or necessity is the presence of particles—solid, semi-solid, or semiliquid matter that has an affinity for water. These particles act as microscopic nuclei upon which water vapor can collect and thereby particle growth is allowed to the mandatory size for cloud droplets and crystals.
- The third prerequisite for cloud formation is a refrigeration mechanism that may lower the temperature of the air and hence raise the relative humidity to saturation.

Clouds are formed when air contains as much water vapor/gas as it can hold. This is called the saturation point, and it may be reached in two ways:

- Moisture adds up until it reaches the maximum amount the volume of air can hold.
- The other approach reduces the temperature of the moisture-filled air, which in turn lowers the amount of moisture it can contain.

Saturation is reached through evaporation and condensation, in that order. Once saturation occurs, moisture becomes visible water droplets in the form of fog and clouds.

The four main ways the clouds can form are:

- Surface heating
- Mountains and terrain
- Air masses being forced to rise
- Weather fronts (cold or warm)

4.6.1 Clouds Formed Due to Surface Heating

First, the sun's rays heat up the ground, which then heats the air. This warm air is lighter and less dense than the surrounding air and hence it begins to rise. When this air rises, it expands and cools and the water vapor condense out of the cool air to produce a cloud. In due course, enough moisture will condense out of the air parcel to form a cloud, as shown in Fig. 4.2. The types of clouds that are formed from the process of surface heating are *cumulonimbus* (and *mammatus* clouds associated with it), *cumulus*, and *stratocumulus*.

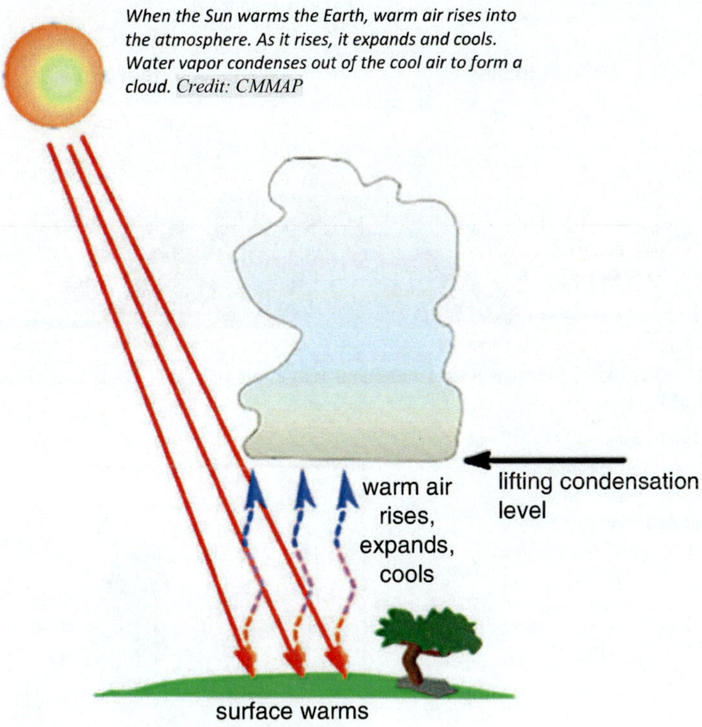

Fig. 4.2 Process of cloud formation

On the contrary, the cooling of the earth's surface leads to the development of fog and stratus clouds.

4.6.2 Clouds Formed Due to Mountains

Some clouds are formed when air encounters a mountain range or other types of terrain. When this takes place, the air will rise and cool, and this cooler air is no longer able to hold all of the water vapor it was able to hold when it was warm. The added water vapor begins to condense out of the air parcel in the form of liquid water droplets and a formation of cloud occurs, as seen in Fig. 4.3. The types of clouds that are formed from the encounters with mountains are *stratus* clouds and *lenticular* clouds.

The other way that mountains cause cloud formation is when air rises because the mountain is warmer than the surrounding air and causes the air to rise. Once the air rises, it follows the same pathway and the process for cloud formation as described above. The types of clouds that are formed in this case are *cumulonimbus* clouds (and *mammatus* clouds associated with it) and *cumulus* clouds.

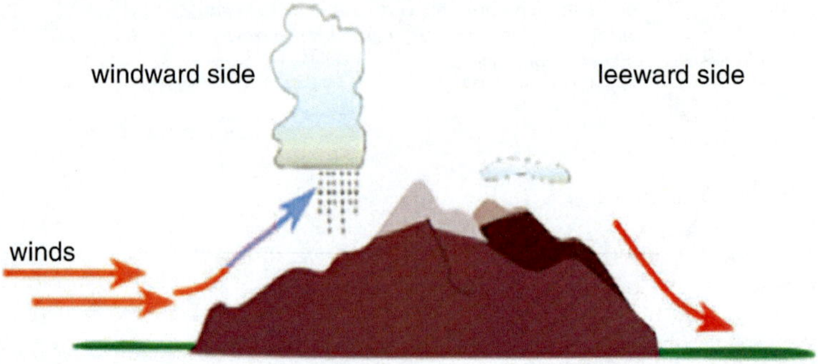

Fig. 4.3 When wind blows across a mountain range, air rises and cools and clouds can form. Credit: CMMAP

Fig. 4.4 Air moves into a low-pressure system and it pushes any air that was there upward. Credit: Windows to the Universe

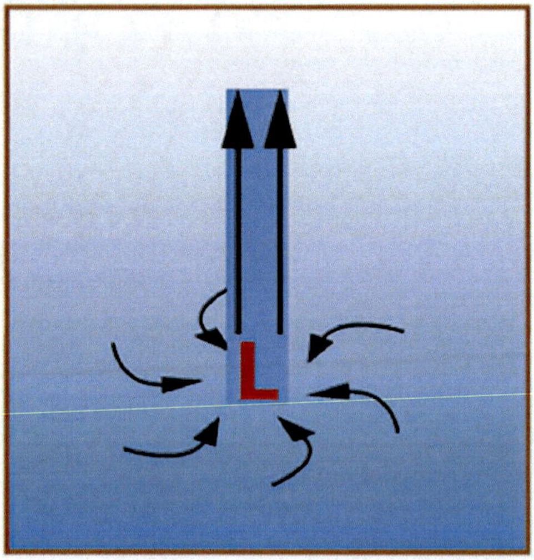

4.6.3 Cloud Formation When Air Is Risen by Force

Some clouds are formed when air at the surface of the earth is forced to rise. There are three processes that force the air to rise. First, in a low-pressure system, air moves in toward the center from all directions because the air moves from high to low pressure. When this air meets in the center, there is nowhere for the air to go but upward. The air is also forced to rise when it moves over or above the land that slopes upward. The air cools as it rises, and in due course clouds will form. Finally, the weather fronts produce clouds by causing air to rise when the lighter warm air flows or travels over the heavier cold air, as seen in Fig. 4.4. All of the cloud types are produced or formed by these processes, especially altocumulus, altostratus, cirrocumulus, stratocumulus, and stratus clouds.

4.6.4 Clouds Formed Due to Weather Fronts

Weather fronts can also cause clouds to form. Fronts appear or occur when two large masses of air collide at the earth's surface.

Warm fronts produce clouds when the warm air replaces the cold air by sliding above it. The following cloud types can be formed in this way: *altocumulus*, *altostratus*, *cirrocumulus*, *cirrostratus*, *cirrus*, *cumulonimbus* (and associated *mammatus* clouds), *nimbostratus*, *stratus*, and *stratocumulus*.

Cold fronts appear or occur when heavy cold air displaces lighter warm air, pushing it upward. Cumulus clouds are the most common cloud types that are formed by cold fronts. Figure 4.5 shows the formation of clouds by weather fronts. They often grow into cumulonimbus clouds, which produce thunderstorms. Cold fronts can also produce the nimbostratus, stratocumulus, and stratus clouds.

4.7 Classification of Clouds

1. *A scheme of distinguishing and grouping clouds according to their appearance and, where possible, their process of formation*
   The one in general use, based on a classification system introduced by Luke Howard in 1803, is that adopted by the World Meteorological Organization and

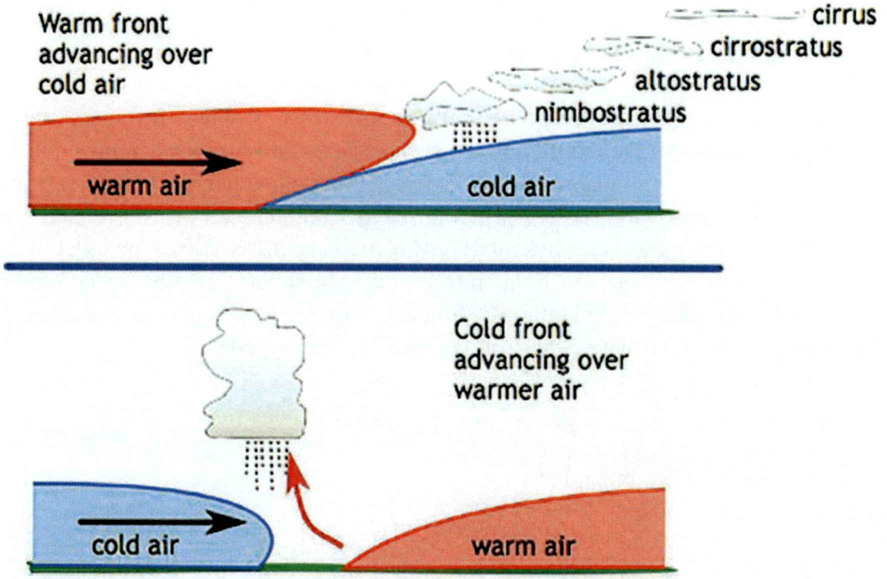

Fig. 4.5 Cold air is denser than warm air, so when a warm air mass meets a cold air mass, the cold air ends up below the warm air. Once the air has risen, it cools and clouds can form. Credit: CMMAP

published in the International Cloud Atlas (1956). This classification is based on the determination of:

(a) Genera—the main characteristic forms of the clouds;
(b) Species—the peculiarities in shape and differences in the internal structure of clouds;
(c) Varieties—special characteristics of arrangement and transparency of clouds;
(d) Supplementary features and accessory clouds—appended and associated minor cloud forms; and
(e) Mother clouds—the origin of clouds, if formed from other clouds.

The ten cloud genera are cirrus, cirrocumulus, cirrostratus, altocumulus, altostratus, nimbostratus, stratocumulus, stratus, cumulus, and cumulonimbus. The 14 cloud species are fibratus, uncinus, spissatus, castellanus, floccus, stratiform, nebulosus, lenticularis, fractus, humilis, mediocris, congestus, calvus, and capillatus. The nine cloud varieties are intortus, vertebratus, undulatus, radiatus, lacunosus, duplicatus, translucidus, perlucidus, and opacus. The nine supplementary features and accessory clouds are incus, mamma, virga, praecipitatio, arcus, tuba, pileus, velum, and pannus.

2. *A scheme of classifying clouds according to their usual altitudes*

   According to their altitudes, three classes are distinguished as high, middle, and low. The high clouds include cirrus, cirrocumulus, cirrostratus, occasionally altostratus, and the tops of cumulonimbus. The middle clouds are altocumulus, altostratus, nimbostratus, and portions of cumulus and cumulonimbus. The low clouds are stratocumulus, stratus, and mostly cumulus and cumulonimbus bases, and sometimes nimbostratus.

3. *A scheme of classifying clouds according to their particulate composition, namely, water clouds, ice-crystal clouds, and mixed clouds*

   The first clouds are composed entirely of water droplets (ordinary and/or supercooled), the second clouds are composed entirely of ice crystals, and the third clouds, a combination of the first two. Of the cloud genera, only cirrostratus and cirrus are always ice-crystal clouds; cirrocumulus can also be mixed; and only cumulonimbus is always mixed. Altostratus is nearly always mixed, but can occasionally be water. All the rest of the genera are usually water clouds, occasionally mixed: altocumulus, cumulus, nimbostratus, and stratocumulus.

4. *A scheme of classifying clouds on the basis of appearance*

   (a) Cirro—curl of hair
   (b) Alto—mid
   (c) Strato—layer
   (d) Nimbo—rain, precipitation
   (e) Cumulo—heap

4.7.1 WMO Cloud Classification

The World Meteorological Organization (WMO) has classified the clouds into ten categories according to their height and appearance. From the height, clouds are grouped into four categories (i.e., family A, B, C, and D) as stated below and there are subcategories in each of these main categories.

The four clouds families, which are in different heights of the troposphere, are:

- High-level clouds (5–13 km)
- Medium-level clouds (2–7 km)
- Low-level clouds (0–2 km)
- Clouds with large vertical extending (0–13 km)

4.7.1.1 High-Level Clouds

High-level clouds occur above about 20,000 ft. and are given the prefix "cirro." Due to cold tropospheric temperatures at these levels, the clouds primarily are composed of ice-crystals and often appear thin, streaky, and white. The three main types of high-level clouds are cirrus, cirrostratus, and cirrocumulus.

1. Cirrus
   Detached clouds in the form of white, delicate filaments, mostly white patches or narrow bands. They have a fibrous (hair-like) and/or silky sheen appearance, as shown in Fig. 4.6a.
2. Cirrostratus
   Transparent, whitish veil clouds with a fibrous (hair-like) or smooth appearance. A sheet of cirrostratus, which is very extensive, nearly always ends up covering the entire sky.
3. Cirrocumulus
   These are thin, white patch, sheet, or layers of clouds without shading. They are composed of very small elements in the form of more or less regularly arranged grains or ripples.

Fig. 4.6 (**a**) Cirrus clouds. (**b**) Cumulus clouds. (**c**) Stratus clouds

4.7.1.2 Medium-Level Clouds

The bases of clouds in the middle level of the troposphere, denoted by the prefix "alto," appear between 6500 and 20,000 ft. Depending on the altitude, time of year, and vertical temperature structure of the troposphere, these clouds may be composed of liquid water droplets, ice crystals, or a combination of the two, including supercooled droplets (i.e., liquid droplets whose temperatures are below freezing). The two main types of middle-level clouds are altostratus and altocumulus.

1. Altostratus
   Gray or bluish cloud sheets or layers of striated or fibrous clouds that totally or partially covers the sky. They are thin enough to regularly reveal the sun as if seen through ground glass.
2. Altocumulus
   White and/or gray patch, sheet, or layered clouds, generally composed of laminae (plates), rounded masses, or rolls. They may be partly fibrous or diffuse.

4.7.1.3 Low-Level Clouds

Low-level clouds are not given a prefix, although their names are derived from "strato" or "cumulo," depending on their characteristics. Low-level clouds occur below 6500 ft. and normally consist of liquid water droplets or even supercooled droplets, except during cold winter storms when ice crystals (and snow) comprise much of the clouds. The two main types of low-level clouds are stratus, which develops horizontally, and cumulus, which develops vertically.

1. Cumulus
   Detached, generally dense clouds with sharp outlines that develop vertically in the form of rising mounds, domes, or towers with bulging upper parts that resemble a cauliflower, as shown in Fig. 4.6b. The sunlit parts of these clouds are mostly brilliant, while their bases are relatively dark and horizontal.
2. Nimbostratus
   These are the continuous rain clouds resulting from thickening altostratus. This is a dark gray cloud layer diffused by falling rain or snow. It is thick enough all throughout to b loc out the sun. The cloud base lowers into the low level of clouds as precipitation continues.
3. Stratus
   A generally gray cloud layer with a uniform base which may, if thick enough, produce drizzle, ice prisms, or snow grains. When the sun is visible through this cloud, its outline is clearly discernible, as shown in Fig. 4.6c.
4. Cumulonimbus
   The thunderstorm cloud, which is a thick and dense cloud in the form of a mountain or huge tower. The upper portion is usually smoothened, fibrous, or striated and nearly always flattened in the shape of an anvil or vast plume.

4.8 Precipitation

Water that falls from the sky is called precipitation. Precipitation forms in clouds. When the air is dry, precipitation is unlikely. However, if there is enough water vapor clouds thicken, rain, snow, and hail begin to fall. Precipitation happens in all seasons. It happens at variety of temperatures, which is why precipitation takes variety of forms. The quantity and distribution decides the nature of vegetation and the nature of cultivated crops. The crop regions are classified on the basis of average rainfall, which is given in Table 4.1.

Precipitation is the total supply of all forms of moisture arising (coming) from the clouds and falling to the ground. It is the most important phase in the hydrological cycle. Precipitation is the immediate source of stream runoff, hence its occurrence, distribution, and intensity determine the hydrologic behavior of streams, namely total discharge, discharge regime, and quality of the discharge. The mechanism of how precipitation is formed in the high atmosphere is a subject matter of "meteorology." Precipitation falls from clouds when already existing cloud elements—droplets or ice crystals—grow large enough to fall at a considerable velocity. The way clouds develop and are structured determines the type of precipitation produced. Steady precipitation of moderate intensity is produced by nimbostratus clouds, downpours originate in cumulonimbus clouds, and drizzle comes from stratus clouds. The types of precipitation that form on the earth's surface are called terrestrial hydrometeors and include dew, rime, and hoarfrost. Atmospheric precipitation is one of the links in the hydrologic cycle of the earth. The amount of precipitation that occurs over several years, in an average month, in a season, or in a year is a determinant of climate and is thus important to a country's agriculture and other sectors of the economy. Other determinants are the distribution of precipitation over the earth, annual and daily cycle, frequency, and intensity.

Precipitation is any product of the condensation of atmospheric water vapor that is deposited on the earth's surface.

- Factors affecting precipitation or the amount of atmospheric moisture over a region
- Climate and geography

Table 4.1 Rainfall range for different regions

Rainfall (mm)	Region
Less than 500	Arid
500–750	Semiarid
750–1000	Subhumid
More than 1000	Humid

4.8.1 How Precipitation Occurs

Precipitation takes place when a body of moisture cools sufficiently for it to become saturated in water droplets and ice crystals upon condensation. Air may be cooled in a number of ways, i.e., by meeting of air masses of different temperature or by coming into contact with a cold object such as ground (mountains). The most important mechanism is uplift of air, as air rises, its pressure decreases, it expands, and cools down. Its ability to hold water is reduced to a certain point. The air becomes saturated and condensation occurs. The formation of cloud does not itself result in precipitation as there must be a mechanism to provide a source of inflow of water. In fact, some clouds are associated with dry weather or light rain, while other clouds are indication of heavy rainfall.

There are four conditions that must be present for the formation of precipitation, which are as follows:

- Atmosphere must have moisture.
- Must be sufficient nuclei present to aid condensation.
- Weather conditions must be good for condensation of water vapor to take place.
- The product of condensation must reach the earth.

4.8.2 Factors Affecting Precipitation

Heavy precipitation occurs near the equator and decreases with the increase in the latitude, i.e., toward polar regions. The main source of moisture for precipitation is evaporation from oceans. Therefore, precipitation tends to be heavier near coastlines. Since lifting of air masses is the cause of almost all precipitation, amount and frequency of rain is generally greater on windward side of the mountain. As downslope motion of air results in decease in humidity, the opposing sides of barriers usually experience relatively light precipitation. High amount of precipitation is reported at higher elevations.

Throughout the year:

- High precipitation occurs in humid season, i.e., July and August.
- Low precipitation occurs in dry season, i.e., May and June.
- Pressure variation in a particular area causes precipitation.
- Temperature variation also causes precipitation in a particular area.

4.8.3 Measurement of Precipitation

The amount of precipitation is usually expressed in units of liquid water depth (e.g., inches) that has fallen on a horizontal surface at a given point over a specific time interval. Meteorological stations use rain gauges and pluviographs to measure precipitation; for large areas, radar is used. The amount of precipitation is expressed as mm depth of water, and the intensity is given in mm per minute, hour, or day.

4.8.4 Precipitation Forms

Precipitation includes all forms of moisture that fall to the earth's surface, such as rain, drizzle, snow, and hail. Precipitation is observed and classified based on its form, type, intensity, and character. Precipitation form is the state that the moisture is in: liquid, freezing, or frozen. Liquid precipitation is any precipitation that falls as a liquid and remains liquid after striking an object, such as the earth's surface or the skin of an aircraft. Rain and drizzle are the two forms of liquid precipitation. Freezing precipitation is any precipitation that falls as a liquid and freezes upon contact with an object, such as freezing rain or freezing drizzle. In this form of precipitation, the liquid water may be a supercooled liquid and freeze upon contact with an object, or the water droplet may have an above-freezing temperature and freeze upon contact with an object that has a below-freezing temperature. (Supercooled liquid have a temperature below their normal freezing temperature, but still exist in the liquid state.). Frozen precipitation is any precipitation of water that falls in its solid state, such as snow, hail, or ice pellets. Different forms of precipitation may occur together, such as mixed rain and snow, but such an occurrence is simply a mixture of forms, not a separate form of precipitation.

4.8.5 Types of Precipitation

4.8.5.1 Rain

It is the liquid precipitation that has a water droplet diameter of 0.02 in. (0.5 mm) or larger. Rain comes from nimbostratus clouds or cumulonimbus clouds. These drops begin in the form of solid crystals, but as they move below the clouds, they begin to melt before they reach the ground and thus create the liquid form we know as rain. When the drops are less than 0.5 mm in diameter, they are known as drizzle, or even smaller, mist. If the water droplets freeze upon contact with a surface, the phenomenon is called freezing rain (glaze). Sometimes rain evaporates before it reaches the ground, resulting in virga. Rain is reported in three intensities, which are:

- Light: for rates of fall up to 2.5 mm/h,
- Moderate: for rates of fall from 2.5 to 7.5 mm/h, and
- Heavy: for rates of fall over 7.5 mm/h.

4.8.5.2 Drizzle

It consists of tiny liquid water drops between 0.1 and 0.5 mm diameter. It is also called mist. Drizzle falls from low strata's and rarely exceeds 1 mm/h. Drizzle is normally produced by low stratiform clouds and stratocumulus clouds. Due to small size, drizzle evaporates before reaching the earth. When drizzle freezes on contact with the ground or other objects, it is referred toas or freezing drizzle. Drizzle usually restricts visibility since it is accompanied by fog.

4.8.5.3 Fog Drip

A cloud of condensed water vapor suspended in the air near the surface level of the earth is called fog. When the water droplets become too heavy to remain suspended, the droplets fall to the ground in the form of fog drip. This form of precipitation originates from condensation in the atmosphere.

4.8.5.4 Snow

Snow is the type of precipitation that consists of white or translucent ice crystals. Each snowflake usually has up to 200 ice crystals. They are all six sided, but they come in different shapes and sizes depending on what temperature they were formed at. Warmer conditions tend to favor larger crystal sizes and clusters of crystals. Although the snow more easily falls through subfreezing air, it is possible for the snow to fall even when the air temperature is above freezing. This happens when the snow is falling into very dry air. With the air above freezing, the snowflakes do partially melt, but because of how dry it is, the water immediately evaporates. This evaporation leads to cooling right around the snowflake and allows it to reach the ground as snow. However, snow that falls like this almost never sticks. There is a version of virga produced by snow called fall streaks. Fall streaks are typically seen with high, wispy cirrus clouds.

Types of Snowfall

- **Snow flurries**: Light snow falling for short durations. There is either no accumulation or just light dusting. Figure 4.7 shows the image of snow flurries.
- **Snow showers**: Snow falling at varying intensities for brief periods of time. Some accumulation is possible, as shown in Fig. 4.8.
- **Snow squalls**: Brief, intense snow showers accompanied by strong, gusty winds. Accumulation may be significant, as shown in Fig. 4.9. These are common in the Great Lakes region.

Fig. 4.7 Snow flurries

Fig. 4.8 Snow showers

Fig. 4.9 Snow squalls

Fig. 4.10 Blowing snow

- **Blowing snow**: Wind-driven snow that reduces visibility and causes significant drifting. Blowing snow may be the snow that is falling and/or loose snow on the ground picked up by the wind, as shown in Fig. 4.10.
- **Blizzards**: Winds over 35 mph with snow and blowing snow, reducing visibility to 1/4 mile.
- **Snow pellets/small hail**: White, opaque, round (or occasionally conical) kernels of snow-like consistency, 0.08–0.2 in. in diameter. They are crisp, easily compressible, and may rebound or burst upon striking a hard surface. Snow pellets occur almost exclusively in snow showers.

- **Snow grains**: Very small, white, opaque grains of ice similar in crystal structure to snow. Whereas the crystal structure of snow has very fine, needle-like branches, the crystal structure of snow grains has thicker, denser elements, with the space between hexagonal branching commonly completely filled. Snow grains do not bounce or shatter on hard surfaces. They usually fall in small quantities, mostly from stratus clouds and never as showers.
- **Hail**: A clear to opaque ball of hard ice, ranging in diameter from 1/8 in. or so to 5 in. or larger, that fall from large thunderstorms. Hail forms when liquid water freezes onto a snowflake to create an ice pellet. They are formed in cumulonimbus clouds where both water and frozen ice occur. They are extremely harmful to crops, and thus have been given the nickname "the white plague." Violent thunderstorms have very strong updrafts that are strong enough to hold ice aloft against the pull of gravity. Hail frequently displays a layered appearance of alternate opaque and clear ice. The opaque layers are created when the hail is in the colder section of the cloud, or gets caught in the downdrafts, and the supercooled droplets freeze onto the hail so quickly that tiny air bubbles get trapped, causing the ice to look milky. When the hail falls into the warmer portion of the cloud, or into the warm updrafts, the supercooled droplets freeze slowly enough that the tiny air bubbles have time to escape before the water freezes, resulting in a sheet of clear ice. The National Weather Service defines severe hail as stones greater than 1 in. in diameter, but even smaller stones can cause significant agricultural impacts.
- **Sleet**: It is the name for ice particles that are formed when snow that melted into raindrops refreeze on their fall through the air. These drops are usually 0.5–5 mm in diameter. Sleet forms when a layer of warmer air, between 5000 and 10,000 ft. above the ground, is trapped between two freezing layers of air. The ice pellets are formed as snowflakes fall from the upper, freezing layer of air, melting on their way through the warmer air, and then refreezing as sleet in the lower, freezing layer of air. They are usually clear ice pellets that fall continuously to the ground; however, they bounce when hitting the ground or other surfaces and will cause very dangerous driving conditions. AccuWeather Senior Meteorologist Brett Anderson said, "Actually, sleet can even provide a little bit of traction for drivers, as opposed to the obvious dangers of a solid sheet of ice that forms from freezing rain."
- **Freezing rain**: Freezing rain occurs when the wedge of warm air aloft is much thicker, allowing the raindrop to survive until it comes in contact with the cold ground. A coating of ice forms on whatever the raindrops contact. Freezing rain is by far the most dangerous because it forms a solid sheet of ice, as opposed to sleet that just has small ice pellets, that quickly bounce off of the surface.
- **Glaze**: Glaze results when snow melts into raindrops but then refreeze upon contact with a surface on the ground. Thick layers of this form of ice, each one 1 mm to 2 cm in diameter, have the potential to break off tree branches and power lines. The rain does not freeze until it reaches the ground, and therefore freezes to it and forms to its surface. Like sleet, glaze also creates the most dangerous

driving conditions. Glaze is most often found in places where the temperatures are right around freezing levels.

- **Rime**: It is a white opaque deposit of ice granular mass or less separated by trapped air and formed by rapid freezing of supercooled water drops. Its specific gravity may be as low as 0.2–0.3.
- **Graupel**: Snow pellets, also known as graupel, occur when snow crystals combine with supercooled water droplets. These droplets are able to retain a liquid state in temperatures well-below freezing. When the supercooled droplets make contact with snow crystals, the droplet freezes around the surface of the snow crystal. These snow pellets have the size ranging from 2 to 5 mm in diameter. Their process of formation is known as accretion, which is the growth of a particle (usually ice pellet or snow flake) by collection of supercooled drops that freeze once they contact the precipitation particle. These particles also contribute to the static electricity build up that helps form lightning.
- **Ice crystals**: Ice crystals, also known as diamond dust, refer to a frozen, ground-level cloud. This type of precipitation is interesting in that it occurs with cloudless skies overhead; the meteorologists call this as clear-sky precipitation. Ice crystals are most common on the polar extremes of the earth, in Antarctica and the Arctic. Diamond dust forms when moist, warm air sits above freezing ground-level air. The two layers mix, pulling the water vapor closer to the freezing temperatures near the ground. When the humidity level is high enough, the water vapor freezes and forms ice crystals.

4.9 Artificial Rainmaking

It is the willful manipulation or modification of local weather where it is attempted to artificially induce moisture in the clouds or increase precipitation, especially to overcome weather extremes and to regulate the weather. Rainmaking is also termed as artificial precipitation, artificial rainfall, and pluviculture.

4.9.1 Principle of Artificial Rainmaking

Rainmaking is based on the principle of coagulation. During artificial rain, silver iodide, potassium iodide, and dry ice are used to spread over the clouds and this process is termed as cloud seeding. The spreading of dry ice stimulates the formation of water droplets. These water droplets collide with each other and upon collision, these water droplets coagulate with each other. The coagulation of water droplets cause clouds to become bulky, which then pours down rain onto the earth's surface.

4.9.2 History of Cloud Seeding

Cloud seeding experiments started with the work of a scientist from General Electric named Vincent Schaefer, who discovered that ice crystals can induce precipitation and created the first artificially induced snow and rainfall. Since ice crystals are difficult to transport and spread over an area, silver iodide, a compound with similar properties, was used as a substitute. The experiments continued until the 1970s when the program was shelved because of lack of usable results.

4.9.3 Purpose of Cloud Seeding

Cloud seeding and artificial rainmaking is a technique of weather modification that improves cloud's ability to induce precipitation, be it rain or snow, by adding condensation nuclei artificially to the atmosphere or dry clouds, providing a base for the raindrops or snowflakes to form. Cloud seeding involves the use of water-absorbent materials to boost the formation of clouds and hence precipitation so that there could be increased crop production in areas where there is little water. Artificial means of producing precipitation are acquiring ever-increasing importance. The practice is to seed clouds with reagents that promote the formation of ice crystals in water clouds and the enlargement of crystals in mixed clouds.

Cloud seeding is one of the tools to mitigate the effects of drought. It is defined as a process in which the precipitation is encouraged by injecting artificial condensation nuclei through aircrafts or suitable mechanism with the aim of influencing the amount or type of precipitation from the clouds. The raindrops are several times heavier than cloud droplets. These mechanisms are different for cold and warm clouds.

4.9.4 Importance of Cloud Seeding

Cloud seeding is useful in the following applications:

4.9.4.1 Increasing Precipitation

Increasing precipitation is one of the common applications of cloud seeding and it is possible with both warm and cold clouds. There are two primary methods used to encourage precipitation:

1. Hygroscopic seeding: initiates warm cloud processes, and
2. Glaciogenic seeding: affects cold cloud processes.

Sometimes both techniques may be helpful and in most cases one can be utilized more effectively than the other. In addition, either technique can be applied from the surface (ground-based) or from an aircraft. Weather Modification, Inc. can help us decide which method will be most effective.

4.9.4.2 Increasing Snowfall

Glaciogenic seeding can also be used to increase precipitation from stratiform and orographic clouds. In such cases, seeding may be accomplished through either ground-based or airborne modes. By increasing snow, the resultant water level during spring is also increased. Hence, cloud seeding increases the water availability for municipal, recreational, and environmental interests.

4.9.4.3 Enhancing Rainfall

Enhancing rainfall can be taken as the major aim of cloud seeding. Efforts to increase rainfall during the warm seasons are typically targeted at convective clouds. Theoretically, it is possible to seed such clouds from ground; however, targeting from aircraft is more efficient and accurate. It is usually possible to affect the cloud through releases of a seeding agent in sub-cloud updrafts, or by dropping the seeding agents directly into the upper regions of the clouds. Warm season glaciogenic seeding is typically applied to treat supercooled cumulus congestus clouds, either by releasing the ice-forming (nucleating) seeding agent in the updraft beneath the actively growing cumulus or by dropping the nucleating agent directly into the supercooled cloud top. The seeding agents can produce ice at significantly warmer temperatures than the natural process. This is how glaciogenic seeding gives the treated cloud a head start in producing precipitation. When clouds do not grow tall and cold enough to produce precipitation through the Bergeron process, it may be possible to stimulate precipitation growth by seeding these warm clouds with hygroscopic seeding agents. This approach can be quite successful through stimulation of the warm cloud precipitation processes. Hygroscopic seeding is normally done from aircraft flying in the sub-cloud updrafts in order to affect the initial cloud droplet development that occurs in this zone.

4.9.4.4 Mitigating Hail Damage

Cloud seeding can be used as a tool to help mitigate hail damage and protect crop yields, homes, and other property, thus reducing the economic harm from disastrous storm damage. Since hail is itself ice that is produced only by vigorous convective clouds, it is certain that such clouds are cold enough to be amenable to glaciogenic seeding techniques. Hail develops when excess supercooled liquid water develops within strong updrafts. However, if the excess supercooled liquid water might be induced to freeze into large numbers of small particles rather than much smaller numbers of large particles, the ice that does precipitate may melt during its transit through the warm sub-cloud layer, or if it does not, it will reach the surface as much smaller, less-damaging ice.

4.9.4.5 Dispersing Fog

Another useful application for cloud seeding is the treatment of ground-based clouds, also known as fog. Supercooled fogs, which composed of water droplets at temperatures cold enough to permit ice development, can easily be cleared by glaciogenic seeding. This can be done either from the ground or from airborne application. The choice between the two will depend on characteristics such as local infrastructure, topography, and wind.

4.9.5 Seeding of Cold Clouds

Seeding cold clouds can be achieved by the following two ways:

4.9.5.1 Dry Ice Seeding

- Dry ice (solid carbon dioxide) has certain specific features. It remains as it is at −80 °C and evaporates, but does not melt. Dry ice is heavy and falls rapidly from top of cloud and has no persistent effects due to cloud seeding.
- Aircrafts are commonly used for cloud seeding with dry ice.
- Aircraft flies across the top of a cloud and 0.5–1.0 cm dry ice pellets are released in a steady stream.
- While falling through the cloud, a sheet of ice crystals is formed and from these ice crystals rain occurs.
- This method is not economical as 250 kg of dry ice is required for seeding one cloud. To carry the heavy dry ice over the top of clouds, special aircrafts are required, which is an expensive process.

4.9.5.2 Silver Iodide Seeding

- Minute crystals of silver iodide produced in the form of smoke acts as efficient ice-farming nuclei at temperatures below −5 °C. When these nuclei are produced from the ground generators, these particles are fine enough to diffuse with air currents.
- Silver iodide is the most effective nucleating substance because its atomic arrangement is similar to that of ice. The appropriate procedure for seeding cold clouds would be to release silver iodide smoke into supercooled cloud from an aircraft.
- For seeding cold clouds, silver iodide technique is more useful than dry ice technique, because substantially less silver iodide is required per cloud. There is no necessity to fly to the top of the cloud if area to be covered is large.

4.9.6 Seeding of Warm Clouds

4.9.6.1 Water Drop Technique

Coalescence process is mainly responsible for growth of rain drops in warm cloud. The basic assumption is that the presence of comparatively large water droplets is necessary to initiate the coalescence process. So, water droplets or large hygroscopic nuclei are introduced into the cloud. Water drops of 25 mm are sprayed from aircraft at the rate of 30 gallons per seeding on warm clouds.

4.9.6.2 Common Salt Technique

Common salt is a suitable seeding material for seeding warm clouds. It is used either in the form of 10% solution or solid. A mixture of salt and soap avoids practical problems. The spraying is done by power sprayers and air compressors or even from ground generators. The balloon burst technique is also beneficial. In this case, gunpowder and sodium chloride are arranged to explode near the cloud base, dispersing salt particles.

4.10 Monsoons

The term monsoon is derived from an Arabic word "Mausim" means "Season." There are different concepts to explain Indian monsoons. Of them the "Thermal concept" proposed by Halley in 1636 is of more practical relevance than other concepts such as aerological and Flohins.

4.10.1 Monsoon in India

Indian monsoon is the most prominent of the world's monsoon systems, which primarily affects India and its surrounding water bodies. It blows from the northeast during cooler months and reverses direction to blow from the southwest during the warmest months of the year. This process brings large amounts of rainfall to the region during June and July.

There are two types of distinguished monsoons in India: (1) Southwest Monsoon (SW) and (2) Northeast Monsoon (NE).

4.10.1.1 Southwest Monsoon

1. In summer, the land mass of India heats quickly and develops a strong low-pressure center, particularly over Northwest India during April and it exists up to September.
2. As the pressure over the adjacent oceans is high, a sea-to-land pressure gradient is established. Therefore, the surface air flow is from the high-pressure areas over the oceans toward the low-pressure areas over the heated land.
3. Even though India should have Northeast Monsoon winds throughout the year due to its position in NE trade wind zone, the SW winds predominate because of the low pressure over Ganges and upper India.
4. The air that is attracted into the centers of low pressure from over the oceans is "warm and moist."
5. This monsoon is active from June to September.
6. The rainfall received is 80–90% of the total annual rainfall of India covering all parts.
7. This monsoon enters Kerala on 1 June and reaches the northern most parts of the country by 15 July, as shown in Fig. 4.11a.
8. There are two branches of the Southwest Monsoon:

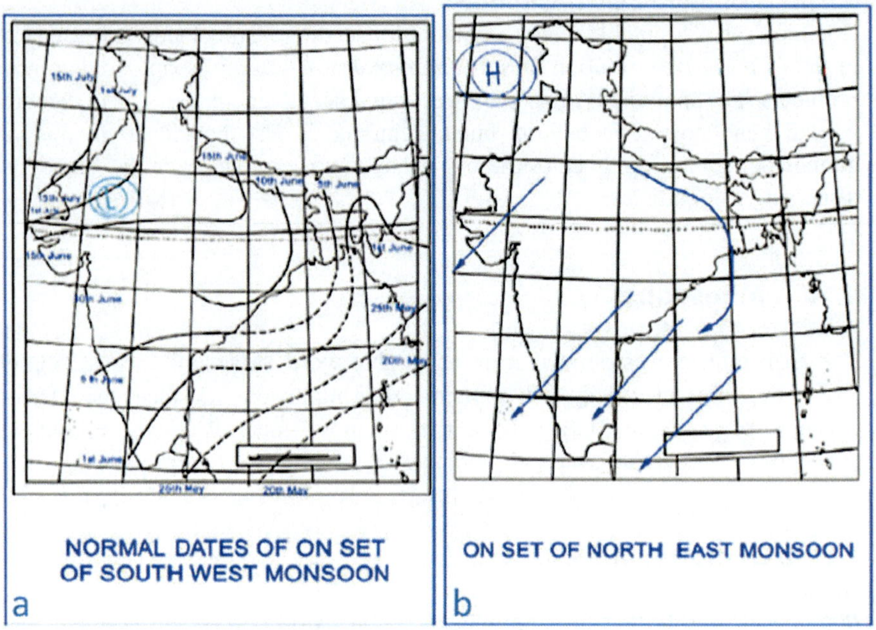

Fig. 4.11 Onset of monsoons in India

(a) The Arabian Sea branch: This branch crosses Western Ghats.
(b) The Bay of Bengal branch: This branch crosses Gangetic Plains.

4.10.1.2 Northeast Monsoon

1. A complete reversal of the Southwest Monsoon winds takes place during winter.
2. In this season the land mass over India cools more rapidly than the surrounding oceans. So, a strong high-pressure center develops over the continent.
3. The pressure over the adjacent oceans is relatively lower. As a consequence, the pressure gradient is directed from land to sea and winds flow in northeast direction. Therefore, there is an outflow of air from the continental land mass to the adjacent oceans.
4. The air flow brings "cold dry" air toward low latitudes, where it receives moisture from the Bay of Bengal that comes across its way.
5. This monsoon is active from October to mid-December.
6. The rainfall received is 10–20% of the total annual rainfall of India, covering parts of Andhra Pradesh (Nellore, Chittoor) and Tamil Nadu, as seen in Fig. 4.11b.
7. The driving mechanisms of monsoon:
   - Differential heating of land and ocean masses causes a pressure gradient and wind is driven accordingly.

- Twist to wind by rotation of the earth.
- Moist process determines strength, vigor, location, and so on.

8. The path of monsoon air is distributed by diverse features such as
   - Earth's rotation,
   - Mountain barriers, and
   - The retarding effect of friction as winds blow overland.

4.10.2 Economic Importance and Influence of Monsoon Rains on Farm Operations

- Nearly 54% of population of the world depends on monsoon for their income.
- Monsoon rains are considered as life-giving rains. Rice or paddy, which is a major food crop, depends on only rainfall for its yield. If rainfall is not uniformly distributed, it results in huge loss of rice crop in particular and all other crops in general.
- Heavy rain during harvesting causes lodging of crop and seed germination. If rainfall does not occur immediately after sowing, it results in germination failure.
- As in the case of other weather elements, the amount and distribution of rainfall influence the crop yield considerably. For example, paddy and sugarcane require high amount of water as compared to groundnut and castor.
  - Timely and evenly distributed rainfall during the crop growth is more beneficial than heavy rainfall all at once.
  - Rainfall of 20 mm is necessary to wet the soil up to a depth of 15 cm, which helps in decomposition of organic matter and also influences the fertility status of the soil by way of leaching of nutrients.
  - Many farm operations such as seedbed preparation, sowing, and intercultivation depend on rainfall.

Bibliography

Ahmed L (2018) Understanding clouds. Research Gate. https://doi.org/10.13140/RG.2.2.35649.86889

Anonymous (2009) Glossary of meteorology

Britannica T (2021) Learn what a climate in geography is. Encyclopedia Britannica. https://www.britannica.com/summary/climate-meteorology

eartheclipse.com (n.d.) https://www.eartheclipse.com/geography/different-types-of-precipitation.html

Geoworld.fandom.com (n.d.) https://geoworld.fandom.com/wiki/fog,hail,mist_and_dew

Globalcryospherewatch.org (n.d.) https://globalcryospherewatch.org/about/about_solidprecip.html

Infobloom (n.d.) https://www.infobloom.com/in-meteorology-what-is-saturation.htm

International cloud atlas (1975) Geneva: secretariat of the world meteorological organization. ISBN: 92-63-10407-7

KBCC meteorology (n.d.) http://kbccmeteorology.pbworks.com/w/page/5507756/forms%20of%20precipitation

Kraght PK (2015) Clouds and cloud formation. Van Nostrand's Scientific Encyclopedia
National Geographic (n.d.) https://www.nationalgeographic.org/encyclopedia/dew/
National Ice and Snow Data Center (n.d.) https://nsidc.org/cryosphere/snow/science/where.html
National Weather Service (n.d.) www.weather.gov
Natural Resources Canada (n.d.) Glaze ice. Archived from the original on 2009-07-15. Retrieved 2009-07-18
NC State University (n.d.) https://climate.ncsu.edu/edu/preciptypes
Pmfias.com (n.d.) https://www.pmfias.com/condensation-forms-of-condensation-types-of-clouds/
Oblack Rachelle (2008) Glaze ice definition. Archived from the original on 2008-08-22. Retrieved 2009-07-18
ResearchGate (n.d.) https://www.researchgate.net/publication/325870303
scied.ucar.edu (n.d.) https://scied.ucar.edu/learning-zone/clouds/how-clouds-form
Shaista M, Bano H, Rather RA, Ahmad S (2018) Cloud seeding; its prospects and concerns in the modern world. J Pure App Biosci 6(5):791–796
Sleet meteorology (n.d.) Encyclopaedia Britannica
Steemit.com (n.d.) https://steemit.com/science/@donfelix/forms-of-precipitation
Tamil Nadu Agricultural University (n.d.-a) http://eagri.org/eagri50/AGRO101/lec25.pdf
Tamil Nadu Agricultural University (n.d.-b) https://www.eagri.org/eagri50/AGRO101/index.html
TNAU Agritech Portal (n.d.) https://agritech.tnau.ac.in/agriculture/agri_agrometeorology_relativehumidity.html
toppr.com (n.d.) https://www.toppr.com/guides/chemistry/environmental-chemistry/condensation/
University of Illinois 2010 Weather world 2010-sleet
usgs.gov (n.d.) https://www.usgs.gov/special-topic/water-science-school/science/condensation-and-water-cycle
Weatheronnline (n.d.) https://www.weatheronline.co.uk/reports/wxfacts/dew.htm
Wikipedia.org (n.d.) https://en.m.wikipedia.org/wiki/dew
Windows to the Universe (n.d.) https://www.windows2universe.org/earth/Atmosphere/clouds/formation_convergence.html
Yourarticlelibrary.com (n.d.) http://www.yourarticlelibrary.com/agrometeorology/condensation-meaning-process-and-types/88791

Microclimate and Climatic Normals

5

Abstract

Microclimate modifications help us to manage and modify extreme weather vulnerability and climatic risks prevailing in crop production or in the nearby vicinity of plants. The main objective of this chapter is to make understand the importance and significance of microclimate modifications in managing the weather extremes and their impact on crop production. Apart from microclimate, the chapter encloses all the dimensions of how climatic normals affect the growth and development of crops. Moreover, how different agroclimatic zones of India contribute significantly to the production of crops has also been discussed in detail.

Keywords

Microclimate · Weather · Climate · Sunlight · Temperature · Agroclimatic regions

5.1 Microclimate

Microclimate is any climatic condition in a relatively small area, within a few meters or less above and below the earth's surface and within canopies of vegetation. The term usually applies to the surfaces of terrestrial and glaciated environments, but it could also pertain to the surfaces of oceans and other bodies of water.

Microclimatic conditions depend on such factors as temperature, humidity, wind and turbulence, dew, frost, heat balance, and evaporation. The effect of soil type on microclimates is considerable. Sandy soils and other coarse, loose, and dry soils, for example, are subject to high maximum and low minimum surface temperatures. The surface reflection characteristics of soils are also important; soils of lighter color reflect more and respond less to daily heating. Another feature of the microclimate is the ability of the soil to absorb and retain moisture, which depends on the

L. Ahmad et al., *Climate Change and Agrometeorology*,
https://doi.org/10.1007/978-981-99-4863-5_5

composition of the soil and its use. Vegetation is an integral part as it controls the flux of water vapor into the air through transpiration. In addition, vegetation can insulate the soil below and reduce temperature variability. Sites of exposed soil then exhibit the greatest temperature variability.

Microclimate, referring to the climatic elements in the immediate vicinity of the plants, is very important as it regulates and determines the plant physiological reactions and energy exchange processes between the plant and its surrounding atmosphere. Lack of optimal climatic elements leads to disturbances in these processes and results in undesired decrease in crop productivity. Microclimatic modifications help in modifying the adverse conditions prevailing in the immediate vicinity of the plants, making it favorable for the better crop growth, development, and yield. Artificial control of field microclimate to maintain the optimum conditions for better plant growth and crop production can be achieved by making field level adjustments. Short-term adjustments at the farm level, such as appropriate sowing time, row spacing and orientation, changes in crop rotation and crop cultivars, changes in soil cultivation and tillage practices, planting method, mulch application, use of shelterbelts/wind breaks, and intercropping, result in the maintenance of favorable crop microclimate by moderating temperature extremes, conserving soil moisture, and increasing radiation interception. Under the present scenario of global warming and increased occurrence of extreme weather events, adoption of such microclimatic modifications in crop production is the need of the hour to manage extreme weather risks and improve crop productivity to attain food security and sustainability of natural resources under changing climatic conditions.

5.2 Processes of Microclimatic Modifications

Microclimatic modification is any process in which all the adverse or stress conditions prevailing in the immediate vicinity of the plants are modified for the better crop growth, development, and yield. Artificial control of plant environment to maintain the optimum conditions for better plant growth and crop production is a future trend in agrometeorological research. Short-term adjustments at the farm level involve production techniques such as changes in crop rotation and crop cultivars, changes in soil cultivation and tillage practices, a shift of sowing dates, adapted fertilization, and crop protection measures.

Long-term adaptations, on the contrary, include major structural changes of farm production systems and need careful agroeconomic planning and realization at societal level; these adaptations also involve a set of sectors and stakeholders such as policy, research, water, and land planning. Some examples of long-term adjustments are changes in land use and landscape structure, breeding and biotechnology applications, crop substitution, and changes in the farm production type. A change in the landscape structure, such as the introduction of windbreaks or hedges, influence the microclimate of crops in neighboring fields, mainly by slowing down the wind speed. Further effects include increased dew formation and leaf wetness duration and a reduction of dry air advection, evapotranspiration, unproductive water loss, and wind erosion.

5.3 Major Techniques of Field Microclimate Modifications

Different crops have specific climatic requirements at different growth stages. Thus, raising crops under favorable microclimate, as per requirements of the crops, needs different management techniques so that sensitive growth stages of the crops can coincide with the favorable microenvironment. Various agronomic manipulations and other microclimate modification options, such as alteration in sowing time, mulch application, row orientation and spacing, irrigation management, planting methods, tillage practices, appropriate cropping systems, and installation of the net houses/screen houses, can be used to make the crop microclimate optimum for highest growth rate and productivity.

5.3.1 Alteration in Sowing Time

Sowing time is an important production component that can be manipulated to counter the adverse effects of environmental stress. This is accomplished through shifting of sowing time so that any stress caused by environment is avoided during the critical stages of crop growth. Matching the phenology of the crop to the duration of favorable environmental conditions by selecting the most appropriate sowing time to avoid the periods of stress is crucial for obtaining maximum yields under changing climate. Favorable microclimate can be created by modifying the date of sowing, which takes care of the deleterious effect of high temperature during reproductive growth period. Temperature is one of the most important elements of climate that determines directly the potential productivity level, particularly for winter crops. Light response, i.e., photoperiodism, which not only control the temperature factor but also the vegetative growth as well as flowering of the plants, is important weather element for wheat crop to assess the thermal response and its requirement at different phenophases to harvest the potential yield. Substantial increase in grain yield of wheat can be achieved by sowing the crop at the optimum time. Heat shock at the end of tillering strongly reduces the rate of leaf photosynthesis, while during grain filling it decreases both rate of photosynthesis and grain growth. The grain yield of cereals has been reported to increase with increase in allocation to the reproductive organs with little increase in biological yield. It would be, therefore, appropriate that plants function in such a manner that maximum amount of dry matter goes to the spikes for increasing weight of grain during post anthesis period, leading to higher grain yield. Partitioning of dry matter at anthesis indicates the maximum contribution from stem, followed by roots, leaves, and spikes, whereas at physiological maturity, the spikes contribute maximum, followed by roots, stem, and leaves. Wheat being a thermosensitive crop, choice of suitable cultivar for different seeding time further gets prime importance.

5.3.2 Mulch Application

Mulch is a layer of material applied at the soil surface, which leads to conserve soil moisture, moderate soil thermal regime, reduce weed growth, and improve fertility and soil health. Mohammad et al. (2012) observed significant increase in grain and straw yield of wheat when the crop residues were retained in the field rather than removed. The increase in water-use efficiency of maize with increase in mulch amount from 2 to 6 Mg/ha. Mulching with crop residues improved water-use efficiency by 10–20% as a result of reduced soil evaporation and increased plant transpiration. Mulch obstructs the solar radiation reaching the soil surface. They increase the soil temperature during winter season and decrease it during summer season. White- and light-colored materials that increase the reflectivity are used to decrease soil temperature during summer season, whereas dark-colored materials and black plastic mulches are used to increase soil temperature during winter season. Straw mulches can be used during both the seasons as they increase soil temperature during winter season and decrease during summer season. Daily mean soil temperature under the plastic mulch is usually 2–9 °C higher than under non-mulching condition. Mulch reduced irrigation water required and evapotranspiration; however, extending mulch duration beyond 60 days had little effect on evapotranspiration.

5.3.3 Row Spacing and Orientation

Row spacing has a great impact on canopy temperature and radiation interception by the crop. If row spacing is more, radiation interception by the crop decreases, more radiation falls on the soil surface and increases the soil temperature. On the other hand, if row spacing is less, radiation interception by the crop increases and its transmission toward soil surface decreases, which decreases the soil temperature. It has been observed that 11% higher grain of wheat was produced in the north–south sowing as compared to the east–west sowing due to its significant effect on crop microclimate. Significantly higher light intensity was observed in north–south direction, whereas canopy temperature was significantly higher in east–west direction.

5.3.4 Irrigation Management

By applying irrigation, soil moisture availability increases, which increases the evapotranspiration and decreases the soil as well as canopy temperature. Under very hot and dry conditions, light irrigation is a common practice to save crops from high temperature stress. Appropriate amount and time of irrigation results in managing heat and water stress and helps to improve yield and water-use efficiency. The increase in irrigations from one to four along with mulch application enhanced intercepted photosynthetically active radiation in potato. In addition, it was noticed that wheat yield and heat-use efficiency increased with increase in irrigation application.

5.3.4.1 Planting Methods

Method of planting also affects the crop microclimate due to its effect on soil properties. The raised beds enhanced 10% yield of wheat due to improved soil physical condition as compared to flat sowing. The higher grain yield and water-use efficiency of chick pea was observed when planted on raised beds as compared to flat planting. The highest yield and water-use efficiency of maize was observed when wheat straw mulch was used along with bed planting.

5.3.4.2 Tillage

Tillage is the physical manipulation of soil, which is intended to destroy weeds, incorporate crop residues and amendments into soil, increase infiltration and reduce evaporation, prepare seed bed, and break hard layer to facilitate root penetration. Tillage practices modify the crop microclimate by affecting the soil properties. Tillage increases the porosity and decreases the thermal conductivity of the soil and increases the soil temperature. The upper loose soil surface also acts as mulch and conserves soil moisture by decreasing evaporation losses.

5.3.4.3 Shelterbelts/Windbreaks

The main function of shelterbelts is to reduce the wind speed and cold advection. Due to reduction in wind speed, they decrease the evapotranspiration and increase the temperature and moisture. Windbreaks modify the microclimate by providing the shelter, which reduces the wind speed and turbulence. Wind breaks protect the plants from the high-intensity winds. They modify the water table by reducing the water losses and hence they improve the crop productivity. Wind breaks also reduce the wind erosion.

5.3.4.4 Cropping Systems

Intercropping, i.e., growing of two or more crops simultaneously on the same field such that the period of overlap is long enough to include the vegetative stage, has significant effect on microclimate and resource-use efficiency.

Intercropping, double cropping, and other mixed cropping practices lead to enhanced efficiency of farm resources with sustainable crop production. Intercropping increases the radiation interception by the canopy and decreases its transmission toward the soil surface, thus decreasing the soil temperature. Improved productivity thus results from either greater interception of solar radiation or higher light-use efficiency or their combination.

5.3.4.5 Protected Cultivation

Protected cultivation makes it possible to obtain increased crop productivity by maintaining a favorable environment for the plants. The use of netting and other type of covering has been shown to restrict air movement around the growing seedlings in higher temperature.

5.4 Agroclimatic Zones of India

The Planning Commission after examining the earlier studies on the regionalization of the agricultural economy has recommended that agricultural planning be done on the basis of agroclimatic regions. For resource development, the country has been broadly divided into 15 agricultural regions based on agroclimatic features, which is shown in Fig. 5.1, particularly soil type, climate such as temperature and rainfall and its variation, and water resources availability as follows:

- Western Himalayan division
- Eastern Himalayan division
- Lower Gangetic Plain region
- Middle Gangetic Plain region
- Upper Gangetic Plain region
- Trans-Gangetic Plain region
- Eastern Plateau and hill region
- Central Plateau and hill region
- Western Plateau and hill region
- Southern Plateau and hill region
- East Coast plain and hill region
- West Coast Plain and hill region
- Gujarat Plain and hill region
- Western Plain and hill region
- Island region

An "Agroclimatic zone" is a land unit in terms of major climates suitable for a certain range of crops and cultivars. The planning aims at scientific management of regional resources to meet the food, fiber, fodder, and fuel wood without adversely affecting the status of natural resources and environment. Agroclimatic conditions mainly refer to soil types, rainfall, temperature, and water availability, which all influences the type of vegetation. An agroecological zone is the land unit carved out of agroclimatic zone superimposed on landform that acts as modifier to climate and length of growing period.

India has been divided into 24 agroclimatic zones by using "Thornthwait indices." Then the Planning Commission as a result of mid-term appraisal of planning targets of VII plan (1985–90) divided the country into 15 broad agroclimatic zones based on physiographic and climate. These are further divided into 72 subzones taking into account the physical attributes and socioeconomic conditions of the region. The emphasis was given on the development of resources and their optimum utilization in a suitable manner within the frame work of resource constraint and potential of each region.

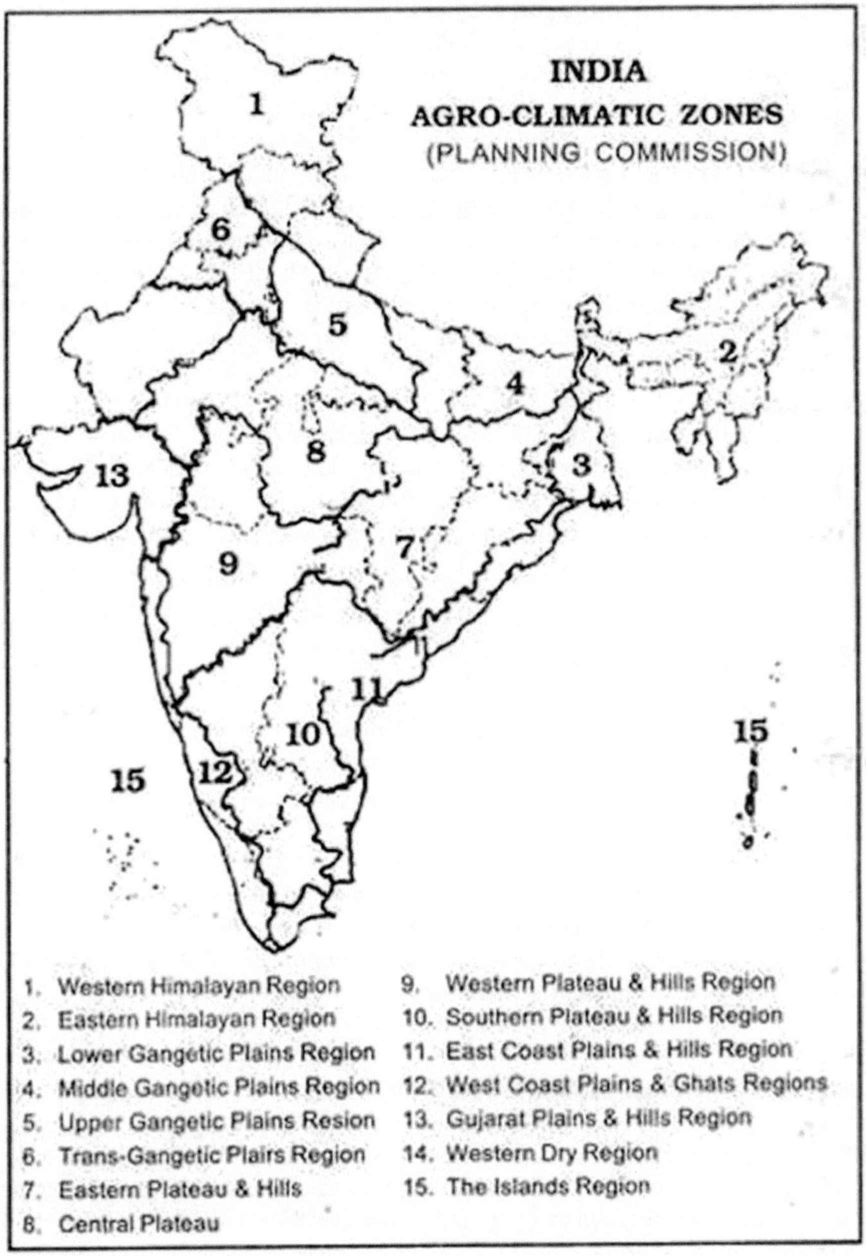

Fig. 5.1 Agroclimatic zones of India

5.4.1 Western Himalayan Region

The Western Himalayan Region covers Jammu and Kashmir, Himachal Pradesh, and the hill region of Uttarakhand. The topography and temperatures show great variation. The average temperature in July ranges between 5 and 30 °C, while in January it ranges between 5 and −5 °C. The mean annual rainfall varies between 75 and 150 cm; in Ladakh, however, it is less than 30 cm. There is alluvial soil in the valleys of Kashmir, Kullu, and Dun, and brown soil in the hills. The valley floors grow rice, while the hilly tracts grow maize in the kharif season. Winter crops are barley, oats, and wheat. The region supports horticulture, especially apple orchards and other temperate fruits such as peaches, apricot, pears, cherry, almond, litchis, and walnut. Saffron is grown in this region. The high-altitude alpine pastures, locally known as "dhoks" or "margs," are used by the Gujjars, Bakarwals, and Gaddis to rear their sheep, goats, cattle, and horses. The economy is largely agrarian. The main problems of this region are poor accessibility, soil erosion, landslides, inclement weather, and inadequacy of marketing and storage facilities. The population is generally rural-based and poor. Research in better seeds and extension service for agricultural development are required.

5.4.2 Eastern Himalayan Region

The Eastern Himalayan Region includes Arunachal Pradesh, the hills of Assam, Sikkim, Meghalaya, Nagaland, Manipur, Mizoram, Tripura, and the Darjeeling district of West Bengal. The topography is rugged. The temperature variation is between 25 and 30 °C in July and between 10 and 20 °C in January. The average rainfall is between 200 and 400 cm. The red-brown soil is not highly productive Jhuming (shifting cultivation) prevails in the hilly areas. The main crops are rice, maize, potato, and tea. There are orchards of pineapple, litchi, oranges, and lime. Infrastructural facilities in the region need to be improved and shifting cultivation is controlled by developing terrace farming.

5.4.3 Lower Gangetic Plain Region

West Bengal (except the hilly areas), eastern Bihar, and the Brahmaputra valley lie in this region. The average annual rainfall lies between 100 and 200 cm. The temperature in July varies from 26 to 41 °C and in January from 9 to 24 °C. The region has adequate storage of ground water with high water table. Rice is the main crop, which at times yields three successive crops (Aman, Aus, and Boro) in a year. Jute, maize, potato, and pulses are other important crops. Planning strategies include improvement in rice farming, horticulture (banana, mango, and citrus fruits), pisciculture, poultry, livestock, forage production, and seed supply.

5.4.4 Middle-Gangetic Plain Region

The Middle-Gangetic Plain region includes large parts of Uttar Pradesh and Bihar. The average temperature in July varies from 26 to 41 °C and that of January 9 to 24 °C, with an average annual rainfall is between 100 and 200 cm. It is a fertile alluvial plain drained by the Ganga and its tributaries. Rice, maize, millets in kharif and wheat, gram, barley, peas, mustard, and potato in rabi are important crops. Alternative farming systems and utilizing chaur lands for pisciculture are some measures to boost agricultural production. Reclamation of user lands, wastelands, and fallow lands for agriculture and allied activities (agroforestry, silviculture, floriculture, etc.) should be done.

5.4.5 Upper Gangetic Plains Region

The Upper Gangetic Plains region includes the central and western parts of Uttar Pradesh and Hardwar and Udham Nagar districts of Uttarakhand. The climate is subhumid continental, with the temperature in July between 26 and 41 °C and in January between 7 and 23 °C. The average annual rainfall is between 75 and 150 cm. The soil is sandy loam. Canal, tube well, and wells are the main sources of irrigation. This is an intensive agricultural region wherein wheat, rice, sugarcane, millets, maize, gram, barley, oilseeds, pulses, and cotton are the main crops. Besides modernizing traditional agriculture, the region needs special focus on dairy development and horticulture. Strategies should include developing multiple mixed cropping patterns.

5.4.6 Trans-Ganga Plains Region

This region (also called the Satluj–Yamuna Plains) extends over Punjab, Haryana, Chandigarh, Delhi, and the Ganganagar district of Rajasthan. Semiarid characteristics prevail over the region, with the mean monthly temperature in July between 25 and 40 °C and that of January between 10 and 20 °C. The average annual rainfall varies between 65 and 125 cm. The soil is alluvial and highly productive. Canals, tube wells, and pumping sets have been installed by the cultivators and the governments. The intensity of agriculture is the highest in the country. Important crops include wheat, sugarcane, cotton, rice, gram, maize, millets, pulses, and oilseeds. The region has the credit of introducing Green Revolution in the country and has adopted modern methods of farming with greater degree of mechanization. The region is also facing the menace of water logging, salinity, alkalinity, soil erosion, and falling water table. Some steps that may be required to make agriculture in the region more sustainable and productive are:

1. Diversion of some rice–wheat area to other crops such as maize, pulses, oilseeds, and fodder;

2. Development of genotypes of rice, maize, and wheat, with inbuilt resistance to pests and diseases;
3. Promotion of horticulture besides pulses such as tur and peas in upland conditions;
4. Cultivation of vegetables in the vicinity of industrial clusters;
5. Supply of quality seeds of vegetables and planting materials for horticulture crops;
6. Development of infrastructure of transit godowns and processing to handle additional fruit and vegetable production;
7. Implementation of policy and programs to increase productivity of milk and wool; and.
8. Development of high-quality fodder crops and animal feed by stepping up area under fodder production.

5.4.7 Eastern Plateau and Hills

This region includes the Chotanagpur Plateau, extending over Jharkhand, Odisha, Chhattisgarh, and Dandakaranya. The region has the advantage of 26–34 °C of temperature in July, 10–27 °C in January, and 80–150 cm of annual rainfall. Soils are red and yellow, with occasional patches of laterites and alluviums. The region is deficient in water resources due to plateau structure and non-perennial streams. Rain-fed agriculture is practiced to grow crops such as rice, millets, maize, oilseeds, ragi, gram, and potato. Steps to improve agricultural productivity and income include cultivation of high-value crops of pulses such as tur, groundnut, and soyabean on upland rain-fed areas, growing of crops such as urad, castor, and groundnut in kharif and mustard and vegetables in irrigated areas; improvement of indigenous breeds of cattle and buffaloes; extension of fruit plantations; renovation including desilting of existing tanks and excavation of new tanks, 95.32 lakh ha of acidic lands through lime treatment; development of inland fisheries in permanent water bodies; and adoption of integrated watershed development approach to conserve soil and rain water.

5.4.8 Central Plateau and Hills

The region is spread over Bundelkhand, Baghelkhand, Bhander Plateau, Malwa Plateau, and Vindhyachal Hills. Semiarid climatic conditions prevail over the region, with the temperature in July between 26 and 40 °C, in January between 7 and 24 °C, and an average annual rainfall from 50 to 100 cm. Soils are mixed red, yellow, and black. There is scarcity of water. Crops grown are millets, wheat, gram, oilseeds, cotton, and sunflower. In order to improve agricultural returns, measures to be adopted are water conservation through water-saving devices such as sprinklers and drip system, dairy development, crop diversification, ground water development, and reclamation of ravine lands.

5.4.9 Western Plateau and Hills

This region comprises southern part of Malwa plateau and Deccan plateau (Maharashtra) and is a region of the regur (black) soil, with the temperature in July between 24 and 41 °C, in January between 6 and 23 °C, and an average annual rainfall of 25–75 cm. Wheat, gram, millets, cotton, pulses, groundnut, and oilseeds are the main crops in the rain-fed areas, while in the irrigated areas, sugarcane, rice, and wheat are cultivated. Also grown are oranges, grapes, and bananas. Attention should be paid to increasing water efficiency by popularizing water saving devices such as sprinklers and drip system. The lower-value crops of jowar, bajra, and rain-fed wheat should give way to high-value oilseeds. Five percent area under rain-fed cotton and jowar could be substituted with fruits such as ber, pomegranate, mango, and guava. Improvement of milk production of cattle and buffalo through cross-breeding along with poultry development should be encouraged.

5.4.10 Southern Plateau and Hills

This region falls in interior Deccan and includes parts of southern Maharashtra, the greater parts of Karnataka, Andhra Pradesh, and Tamil Nadu uplands from Adilabad District in the north to Madurai District in the south. The mean monthly temperature of July varies between 25 and 40 °C and the mean temperature in January is between 10 and 20 °C. The annual rainfall is between 50 and 100 cm. It is an area of dry-zone agriculture where millets, oilseeds, and pulses are grown. Coffee, tea, cardamom, and spices are grown along the hilly slopes of Karnataka plateau. Some of the area under coarse cereals may be diverted to pulses and oilseeds. Horticulture, dairy development, and poultry farming should be encouraged.

5.4.11 Eastern Coastal Plains and Hills

In this region are the Coromandel and northern Circar coasts of Andhra Pradesh and Odisha. The mean temperature in July ranges between 25 and 35 °C and the mean temperature in January varies between 20 and 30 °C. The mean annual rainfall varies between 75 and 150 cm. The soils are alluvial, loam, and clay and are troubled by the problem of alkalinity. The main crops include rice, jute, tobacco, sugarcane, maize, millets, groundnut, and oilseeds. The main agricultural strategies include improvement in the cultivation of spices (pepper and cardamom) and development of fisheries. These involve increasing cropping intensity using water-efficient crops on residual moisture, discouraging growing of rice on marginal lands, and bringing such lands under alternate crops such as oilseeds and pulses, diversifying cropping and avoiding mono-cropping, developing horticulture in upland areas, social forestry, and dairy farming.

5.4.12 Western Coastal Plains and Ghats

Extending over the Malabar and Konkan coastal plains and the Sahyadris, the region is humid, with the mean temperature in July varies between 25 and 30 °C and mean temperatures in January between 18 and 30 °C. The mean annual rainfall is more than 200 cm. The soils are laterite and coastal alluvial. Rice, coconut, oilseeds, sugarcane, millets, pulses, and cotton are the main crops. The region is also famous for its plantation crops and spices, which are grown along the hill slopes of the Western Ghats. The agricultural development must focus attention on production of high-value crops (pulses, spices, and coconut). The development of infrastructural facilities and promotion to prawn culture in brackish water should be encouraged.

5.4.13 Gujarat Plains and Hills

This region includes the hills and plains of Kathiawar and the fertile valleys of Mahi and Sabarmati rivers. It is an arid and semiarid region, with the mean temperature in July reading 30 °C and that of January about 25 °C. The mean annual rainfall varies between 50 and 100 cm. Soils are regur in the plateau region, alluvium in the coastal plains, and red and yellow soils in Jamnagar area. Groundnut, cotton, rice, millets, oilseeds, wheat, and tobacco are the main crops. It is an important oilseed producing region. The main strategy of development in this region should be canal and groundwater management, rain water harvesting and management, dry land farming, agroforestry development, wasteland development, and developing marine fishing and brackish/back-water aquaculture development in coastal zones and river deltas.

5.4.14 Western Dry Region

Extending over Rajasthan, West of the Aravallis, this region has an erratic rainfall of an annual average of less than 25 cm. The desert climate further causes high evaporation and contrasting temperatures – 28–45 °C in June and 5–22 °C in January. Bajra, jowar, and moth are main crops of kharif and wheat and gram of rabi. Livestock contributes greatly in desert ecology. The main areas in need of a thrust for development are rainwater harvesting; increasing yield level of horticultural crops such as water melon, guava, and date palm; adopting high-quality germplasm in cattle to improve their breed; and adopting silvi-pastoral system over wastelands.

5.4.15 Island Region

The island region includes Andaman and Nicobar and Lakshadweep, which have typical equatorial climates (annual rainfall less than 300 cm; the mean July and January temperature of Port Blair being 30 and 25 °C, respectively). The soils vary from sandy along the coast to clayey loam in valleys and lower slopes. The main crops are rice, maize, millets, pulses, areca nut, turmeric, and cassava. Coconut covers nearly half of the cropped area.. The area is covered with thick forests and agriculture is in backward stage. The main thrust in development should be on crop improvement, water management, and fisheries. Improved variety of rice seeds should be popularized so as to enable farmers to take two crops of rice in place of one. For fisheries development, multipurpose fishing vessels for deep sea fishing should be introduced, suitable infrastructure for storage and processing of fish should be built up, and brackish water prawn culture should be promoted in the coastal areas.

5.5 Classification by ICAR

The State Agricultural Universities have divided each state into subzones under the National Agricultural Research Project (NARP) of ICAR. Based on the rainfall pattern, cropping pattern, and administrative units, a total of 127 agroclimatic zones are indicated. The numbers of zones in each state are given in Table 5.1.

5.6 Climatic Normal for Crop Production

5.6.1 Rainfall Probability Analysis

Rainfall is one of the most important factors in crop production planning, particularly under rain-fed and dry land situations. As the monthly, seasonal, and annual rainfall of a region is useful in designing water harvesting structures, the weekly rainfall analysis gives more useful information for crop planning. In high rainfall areas, the distribution of rainfall is of more concern as compared to the amount received. Erratic nature of rainfall intensity and frequency often make crop planning a difficult task in rain-fed areas. From probabilistic analysis, one can find out suitable

Table 5.1 Zones in each state as per ICAR

State	No. of zones	State	No. of zones
Andhra Pradesh	7	Madhya Pradesh	12
Assam	6	Maharashtra	9
Bihar	6	NE hill region	6
Gujarat	8	Odisha	10
Haryana	2	Punjab	5
Himachal Pradesh	4	Rajasthan	9
Jammu and Kashmir	5	Tamil Nadu	7
Karnataka	10	Uttar Pradesh	10
Kerala	5	West Bengal	6
Total 127			

period for harvesting of rainwater and recycling of the same as supplemented/ lifesaving irrigation during lean periods of rainfall. The main assumption in this method is that rainfall is, more or less, normally distributed. The variability of rainfall, which is the ratio of standard deviation to mean of rainfall for a long period for any specific location, is another most important factor that determines crop production in any given place. High rainfall variability in rain-fed area may hamper crop production. Generally, areas with low rainfall are having high rainfall variability. The North Eastern region falls in low rainfall variability category and the coefficient of rainfall variation ranges from 8 to 15%. Still, under the influence of global climate change, even high rainfall areas are facing drought-like situations in recent years and floods are more common in low rainfall areas. The long-term average of rainfall for any specific period such as week, month, season, or year is considered as normal rainfall for that particular place. Monsoon is the major determining factor for availability of water for crop production in the country. It also affects ground water recharge as well as availability of soil moisture for rabi crops. When the amount of monsoon rainfall receipt is within ±19% of the long-term normal value, then it is considered that monsoon was normal in that year. However, erratic distribution and intensity of rainfall are two other factors for crop failure in many parts of the country even though monsoon was found normal in a particular year. For the North Eastern states of India, the normal annual rainfall ranges from 200 to 300 cm.

5.6.2 Productivity Zoning for Crops

Productivity zoning with respect to any particular crop on a regional scale is an important tool to have an elaborate idea about pros and cons of production and productivity of crops, which is derived from blending of long-term area and productivity data of any given crop with existing climatic conditions. It helps in finding out areas with low productivity and possible solutions to do away with the constraints.

5.7 Crop Calendar

A crop calendar is an instrument to rationalize crop activities in relation to weather condition, which is presumed to be happening in a particular period of the year. For convenience, crop calendar is formulated in weekly time frame, indicating most probable temperature and rainfall for that particular week. Depending on the probable weather, the timing for major crop activities such as field preparation, manuring, seed sowing, nursery raising, and harvesting can be adjusted to capitalize on favorable weather condition so as to minimize risk factor involved due to weather on crop production. This has become very handy in increasing cropping intensity by selecting two or more crops in a crop calendar year by adjudging suitability to grow different crops in a given piece of land.

5.8 Soil Water Balance

Soil water balance study is useful in agriculture by various ways. A water balance model is a simple method to calculate crop water use. It can be used in intelligent planning of long-range water resource management. This knowledge can be utilized to modify components of water balance so that water can be utilized best in crop production. Thornthwaite's Book-keeping Water Balance Method is one of the simplest methods to determine soil water balance. For this method, the requirements are the data of mean monthly temperature, latitude of the station, monthly precipitation, and tables and charts, which are available in the book.

5.9 Quantification of Crop Response to Water

Soil, plant, and water relationships involve many complex biological, physiological, physical, and chemical processes. However, from crop production point of view, quantification of such relationships is of principal importance to know crop water requirements, which are supported by decisions such as when and how to irrigate and how much water to apply at the field level, as shown in Table 5.2. For application in planning, design, and operation of irrigation schemes, it is possible to analyze the effect of water supply on crop yields. Water deficits in crops, and the resulting water stress on the crop, have an effect on crop evapotranspiration and crop yield. Water stress in the crop can be quantified by the rate of actual evapotranspiration (ETa) in relation to rate of maximum evapotranspiration (ETm). When crop water requirements are fully met from available water supply, then ETa = ETm; when water supply is insufficient, ETa < ETm. For most crops and climates, ETa and ETm can be quantified. The manner in which water deficit affects crop growth and yield varies with species, varieties, and crop growth period. In order to quantify the effect of water stress, it is necessary to derive the relationship between relative yield decrease and relative evapotranspiration deficit given by the empirically derived yield response factor.

Table 5.2 Decisions for effective crop water requirement

Strategic decisions before the growing season	Tactical decisions at the start and in the course of growing season
Choice of land	Planting time
Drainage	Crop species/cultivars
Crop rotation	Plant density, weeding
Crop variety/varieties	Irrigation
Soil tillage	Top dressings of fertilizer
Types of fertilizers to be used	Pest/disease control, harvesting

Source: Saikia, 2018

$$(1 - \mathrm{Ya/Tx}) = \mathrm{Ky}(1 - \mathrm{ETa/ETx})$$

where Ya is the actual harvested yield, Ym is the maximum harvested yield, and ky is the crop response factor.

Maximum evapotranspiration or potential evapotranspiration can be defined as the amount of evapotranspiration from a short green grass of 1 ft. uniform height grown in a non-stress condition of water supply. Potential evapotranspiration can be easily determined using empirical models developed by different workers after thorough evaluation in any given set of environmental conditions.

Among different empirical models, the Papadakis model fit best for the hilly environment of Meghalaya in the altitude range of 800–1400 m above mean sea level.

The Papadakis model can be expressed as:

$$\mathrm{PET} = 0.5625 \times (e_{\max} - e_{\min - 2}) \times 10$$

where PET is the monthly potential evapotranspiration (mm), $e_{\max}$ is the saturation vapor pressure at mean maximum temperature (mb), and $e_{\min - 2}$ is the saturation vapor pressure at mean maximum temperature (mb) minus 2 °C.

Types of decisions that can be taken based on various agroclimatic analyses are given in Table 5.2.

Bibliography

Ahmad L, Habib Kanth R, Parvaze S, Sheraz Mahdi S (2017) Experimental Agrometeorology: A Practical Manual. Springer, Cham. https://doi.org/10.1007/978-3-319-69185-5

Alexandrov V, Eitzinger J, Cajic V, Oberforster M (2002) Potential impact of climate change on selected agricultural crops in north-eastern Austria. Glob Chang Biol 8(4):372–389

Britannica, T. Editors of Encyclopaedia (2020, March 19) Microclimate. In: Encyclopedia Britannica https://www.britannica.com/science/microclimate

Chen CC, McCarl BA (2001) An investigation of the relationship between pesticide usage and climate change. Clim Chang 50(4):475–487

Cleugh HA (1998) Effects of windbreaks on airflow, microclimates and crop yields. Agrofor Syst 41(1):55–84

Donald CM, Humblin J (1976) The biological yield and harvest index of cereals as agronomic and plant breeding criteria. Adv Agron 28:361–405

Egli DB (2004) Seed-fill duration and yield of grain crops. Adv Agron 83:243

Eitzinger J, Orlandini S, Stefanski R, Naylor REL (2010) Climate change and agriculture: introductory editorial. J Agric Sci 148(5):499–500

Fahong W, Xuging W, Sayre K (2004) Comparison of conventional flood irrigated flat planting with furrow irrigated raised bed planting for winter wheat in China. Field Crop Res 87:35–42

Gebru H (2015) A review on the comparative advantages of intercropping to mono-cropping system. J Biol Agric Healthcare 5(9):1–13

Ghaffari AHFC, Cook HF, Lee HC (2002) Climate change and winter wheat management: a modelling scenario for south-eastern England. Clim Chang 55(4):509–533

Gomez AA, Gomez KA (1983) Multiple cropping in the humid tropics of Asia. International development research Centre, Ottawa

Imran S, Arif M, Khan A, Khan MA, Shah W, Latif A (2015) Effect of nitrogen levels and plant population on yield and yield components of maize. Adv Crop Sci Technol 3:1–7

Iqbal MM, Akhter J, Mohammad W, Shah SM, Nawaz H, Mahmood K (2005) Effect of tillage and fertilizer levels on wheat yield, nitrogen uptake and their correlation with carbon isotope discrimination under rainfed conditions in northwest Pakistan. Soil Tillage Res 80:47–57

Jalota SK, Prihar SS (1990) Bare-soil evaporation in relation to tillage. In: Advances in Soil Science 12. Springer, New York, NY, pp 187–216

Kar G, Kumar A (2007) Effects of irrigation and straw mulch on water use and tuber yield of potato in eastern India. Agric Water Manag 94(1–3):109–116

Kingra PK, Kaur P (2012) Effect of dates of sowing on thermal utilisation and heat use efficiency of groundnut cultivars in central Punjab. J Agricult Phys 12(1):54–62

Krishnan A, Singh M (1972) Soil-climatic Zones in Relation to Cropping Patterns. In: Proceedings of Symposium on Cropping Patterns in India. ICAR, New Delhi

Ministry of Jalsakthi (n.d.) http://jalshakti-dowr.gov.in/agro-climatic-zones

Muhammad I, Khan F, Khan A, Wang J (2018) Soil fertility in response to urea and farmyard manure incorporation under different tillage systems in Peshawar, Pakistan. Int J Agric Biol 20: 1539–1547

National Research Council (1993) Sustainable agriculture and the environment in the humid tropics. National Academies Press, Washington

Olesen JE, Trnka M, Kersebaum KC, Skjelvåg AO, Seguin B, Peltonen-Sainio P, Rossi F, Kozyra J, Micale F (2011) Impacts and adaptation of European crop production systems to climate change. Eur J Agron 34(2):96–112

Pal R, Mandal D, Naik BS (2017) Effect of different meteorological parameters on the development and progression of rice leaf blast disease in western Odisha. Int J Plant Protect 10(1):52–57

Pandey GK, Singh HS, Bharti PK and Meshram MK (2013) Design of stepped monopole UWB antenna with WLAN band notched using modified mushroom type EBG structure. In 2013 IEEE International Conference on Electronics, Computing and Communication Technologies (pp. 1-6). IEEE

Saika US, Goswami B (2018) Agroclimate and crop planning. In: Das KMA (ed) Conservation Agriculture for advancing food security in changing climate, vol 1. Today & Tomorrow's Printers and Publishers, ICAR Research Complex for NEH Region, Umaiam, Meghalaya, pp 155–167

Schapendonk AHCM, Xu HY, Van Der Putten PEL, Spiertz JHJ (2007) Heat-shock effects on photosynthesis and sink-source dynamics in wheat (*Triticum aestivum* L.). NJAS-Wageningen J Life Sci 55(1):37–54

Singh AK (2008) Heat unit requirements for phenophases of wheat genotypes as influenced by sowing dates. J Agrometeorol 10(2):209–212

Singh P, Nedumaran S, Boote KJ, Gaur PM, Srinivas K, Bantilan MCS (2014) Climate change impacts and potential benefits of drought and heat tolerance in chickpea in South Asia and East Africa. Eur J Agron 52:123–137

Singh MC, Yousuf A, Singh JP (2016) Greenhouse microclimate modeling under cropped conditions: a review. Res Environ Life Sci 9:1552–1557

Tolera T (2003) The Conditions of Kebele Managed Rental Accommodations in Addis Ababa: A Comparative Study of Woredas 5 And 26. Doctoral dissertation, Addis Ababa University

Trnka M, Dubrovský M, Semeradova D, Žalud Z (2004) Projections of uncertainties in climate change scenarios into expected winter wheat yields. Theor Appl Climatol 77(3):229–249

Tubiello FN, Donatelli M, Rosenzweig C, Stockle CO (2000) Effects of climate change and elevated CO2 on cropping systems: model predictions at two Italian locations. Eur J Agron 13(2–3):179–189

Tyagi PK, Pannu RK, Sharma KD, Chaudhary BD, Singh DP (2004) Post-anthesis dry-matter accumulation and its partitioning in different wheat (Triticum aestivum) genotypes under varying growing environments. Indian J Agronomy 49(3):163–167

Willey RW (1990) Resource use in intercropping systems. Agric Water Manag 17(1–3):215–231

Yang W, Peng S, Dionisio-Sese ML, Laza RC, Visperas RM (2008) Grain filling duration, a crucial determinant of genotypic variation of grain yield in field-grown tropical irrigated rice. Field Crop Res 105(3):221–227

Weather Disasters and Management, Air Pollution, and Role of Meteorology

6

Abstract

Weather disasters are the extreme agrometeorological events that affect not only vegetation but subsequently affect the population and livestock as well. There is a close relationship between disaster severity and the consequent impact on agriculture and postproduction. In this chapter, we will try to analyze the use of technology in mitigation of air pollution and other different types of weather disasters posing a potential threat to the agricultural systems.

Keywords

Weather disasters · Air pollution · Flood · Drought · Hails · Frost · Heat wave · Cold wave · Cyclones · Anticyclones · Thunderstorms

6.1 Disaster

Disaster is an event, natural or man-made, sudden or progressive, which impacts with such severity that the affected community has to respond by taking exceptional measures (W. Nick Carter). Most of the world's natural disasters occur in Asia and the Pacific. Environmental degradation can be seen as a result of economic development and associated improper human settlement patterns that increase a country's vulnerability to natural hazards and aggravate the impacts. Disasters tend to disrupt the normal patterns of life, which are usually sudden, unexpected, severe, devastating, and widespread. The occurrence of natural calamities such as droughts, floods, cyclones, and heat waves has increased in India in the recent past. The trend of precipitation is shifting toward an increase of heavy precipitation events and long dry spells.

An extreme agrometeorological event is the interaction between a vulnerable agricultural system and extreme weather conditions. However, it includes weather

L. Ahmad et al., *Climate Change and Agrometeorology*,
https://doi.org/10.1007/978-981-99-4863-5_6

conditions conducive to the development of agents such as pests and diseases that adversely affect all aspects of agriculture, including livestock and pasture, forests, and fisheries.

6.2 Types of Extreme Agricultural Events

The extreme climatic events affect agricultural production by altering the normal values of the climatic variables that are required for normal growth and development of a plant. Generally, a plant exhibits a normal threshold value for each of these variables for its effective growth and development. Change in any of these variables cause deviation from normal physiological processes within the plant. Some of the main extreme climatic events from agricultural point of view are:

(a) Tropical storms (cyclones, hurricanes, typhoons, etc.) associated with high winds, flooding, and storm surges;
(b) Floods (other than those related to tropical storms) and heavy rains during monsoon and water logging;
(c) Thunderstorms and hailstorms;
(d) Drought;
(e) Dust storms and sand storms;
(f) Low temperature, frost, and snow;
(g) Weather conducive to fires (lightning); and
(h) Weather encouraging pests and diseases of crops and livestock.

Besides these agrometeorological events, we also encounter many other geological or geophysical events that can cause considerable loss to the property and lives and these are:

(a) Earthquakes and volcanic eruptions
(b) Tsunami
(c) Avalanche
(d) Landslides and mudslides

6.2.1 Tropical Storms

Tropical storms occur due to an interaction between ocean and atmosphere. According to the World Meteorological Organization (WMO) data survey, "About 70 per cent of tropical disturbances which later develop into tropical storms, form in the Northern Hemisphere." The radial dimensions of tropical storms vary from 50–100 km radius to 2000 km radius. Tropical storms are a low-pressure system that operates in anticlockwise direction in Northern Hemisphere and clockwise direction in Southern Hemisphere. It is called by different names at different places: in Atlantic and Eastern Pacific, it is known as Hurricanes; in Western Pacific,

Typhoons; in Australia, willy-willies; in Mexico, Cordonazo; in Philippines, Bagious; and in Indian seas it is known as Cyclones.

They are formed in the ocean where the surface temperature is above 27 °C. Tropical storms or cyclones are associated with winds that often exceed 200 km/h; rainfall that exceeds 5–100 cm in 24 h, and worst of all, very high storm tides (storm surge combined with astronomical tides) that often exceed 6 m. Cyclones consist of two main parts: the first one is the center portion referred to as "Eye" of the cyclone, which is intensely low-pressure point where low winds blow, and the second one consists of "Eye wall," which is a high-pressure zone where high-speed winds blow. It is the most devastating phenomenon for the people of coastal areas, resulting in loss of lives, loss of assets, and total disruption and destruction of crops. The average life period of cyclone in North Indian Ocean is 2–5 days. In India there are two main cyclone seasons that operate from Indian seas: pre-monsoon cyclone season: March to May; post-monsoon cyclone season: October to December.

6.2.2 Anticyclones

Anticyclones have a flow opposite to that of cyclones – i.e., an outward-spiraling motion, with the winds rotating clockwise in the Northern Hemisphere and counter-clockwise in the Southern Hemisphere. They are usually not as strong as the cyclonic storms and bring no rainfall.

6.2.2.1 Effects of Cyclone on Agriculture

The agricultural sector includes field crops, horticultural crops, agroforestry plants, livestock, and fisheries. Cyclones in coastal areas severely affect all these components of agriculture sector through direct damage by high-speed wind, torrential rain, and extensive flooding. High tide may bring in saline water and sand mass, making the fields unsuitable for agriculture. The indirect effects include infection and disease of farm animals, fish, and crop plants. Agricultural marketing and trade are adversely affected due to lean season of animal, fish, and crop production. Super cyclone in 1999 and Phailin in 2013 severely affected crop production and livelihood of farmers in coastal areas in Eastern Coast of India.

Management and Mitigation

For any effective cyclone management, public awareness plays the most important role in order to prevent the damages. The agriculture gets badly hit during and after cyclonic event. Hence there should be efficient crop management practices to either reduce or escape the cyclonic events. The crops should be harvested early if matured, and the harvest should be safely stored. This is referred to as "Preparedness" in agriculture system. Irrigation canals and embankments of rivers in the risk zone should be repaired to avoid breaching. An effective wind break plantation system along the coastal areas helps in reducing the speed of winds to a greater extent. Risk zone mapping and analysis of land use pattern should be undertaken to guide growth and development away from cyclone-prone areas. Beyond this, as the storm approaches the area, nothing more can be done. Cyclones cannot be prevented but an increase in public awareness and effective pre-cyclone, on-cyclone, and post-cyclone measures can definitely reduce their potential disaster to a very large extent.

The mitigation processes include banana production technology, rice production technology, and nursery management practices. Banana is a water-loving plant and loves moist conditions, hence it is more suitable for coastal areas. Similarly, rice grows indefinitely well in anaerobic conditions. Moreover, we can use bio-drainage plants that decrease the water table by absorbing huge amounts of water and making the soil appropriate for cultivation.

6.2.3 Heavy Rainfall, Floods, and Inundation

Heavy rainfall is abnormally large amounts of liquid precipitation and is caused due to many phenomena such as low-pressure systems, upper air cyclonic circulations, and thunderstorms/hailstorms. The spell or intensity of rain rate varies from time to time, place to place, and even hour to hour. A spell of 4–16 mm/h is considered to be as a Heavy spell; Very Heavy spell is the rain rate of 16–50 mm/h, and the Extremely Heavy spell can be considered when the rain rate is 50–100 mm/h. If the rain rate exceeds 100 mm/h, it is referred to as cloudburst. Heavy rainfall often causes heavy runoffs, flash floods, and inundations, all of which consequently damages agri/horticulture production. In 2008 and 2014, the Ladakh and Kashmir regions of Jammu and Kashmir experienced catastrophic flash floods.

6.2.4 Floods

Floods are the most common and widespread of all natural disasters. India is one of the highly flood prone countries in the world. As per the National Flood Commission report, around 40 million hectares of land in India is prone to floods. Floods are the consequence of heavy rainfall and water logging. The intensity of floods depends on factors such as terrain, soil characteristics, and available drainage system, and it gets worsened by continuous rainfall. If the terrain is bare and sloppy, then the water absorbing capacity is very low, increasing the potential runoff and erosion. However, if the soil surface is accumulating and low lying, the absorbing capacity of the soil increases manifold. Inundation of the soil also depends on the prior availability of moisture. If the soil is dry, then it can absorb much of the rainfall and if the soil conditions are moist then absorbing capacity is reduced.

6.2.4.1 Effect of Flood on Agriculture

According to Gommes and Negre, 1992; The effects on agriculture can be classified as direct or indirect. According to them, direct effects are those that affect the property and income of individuals, enterprises, and the public sector. An example would be the loss of a current crop of maize due to severe flooding, while indirect effects are slower and often more widespread (geographically, economically, etc.)

than direct effects and result from decreased income, environmental degradation, and other factors. The processes that come under direct and indirect effects are mentioned as follows:

Fallow period

- Loss of topsoil
- Loss of soil nutrients
- Soil compaction
- Soil erosion
- Deposition of undesirable materials
- Permanent damage to perennial crops, trees, livestock, buildings, and machinery.
- Anaerobic processes
- Permanent cessation of farming in floodplains
- Permanent diversion/realignment of rivers, streams, other bodies of water, and settlements
- Loss of livestock and/or habitat

Growing season

- Water logging of crops
- Lodging of standing crops
- Loss of soil nutrients
- Soil erosion
- Greater susceptibility to diseases and insects
- Interruptions to tillage, planting, crop management, and harvesting
- Permanent damage to perennial crops, trees, livestock, buildings, and machinery
- Soil temperature disturbance
- Transportation interruptions
- Grain spoilage, in-field and off-site

6.2.4.2 Management and Mitigation

The efficient management practices include the thorough documentation of past events without which it is difficult to anticipate future conditions. The documentation helps us to find out the frequency of occurrence of the floods and the time gap in which it can recur. For heavy rainfall documentation, hourly (good), 15-min (better), or breakpoint (best) observations are needed. There should be efficient and reliable instrumentation for measuring rainfall, other types and forms of precipitation (including snow water equivalent), stream flow, lake and reservoir levels, and soil moisture and temperature, e.g., rain gauge. There should be effective use of radar and satellite-derived fields for estimation of precipitation. The ground-based preparations such as dredging of rivers, streams, prevention illegal constructions on lake sides, bunds, and so on are necessarily important to accommodate water flow in times of heavy rainfall. NDRF preparedness finally plays an important role once the incident has taken place.

6.2.5 Thunderstorm and Hailstorm

Thunderstorm and hailstorm occur due to Cumulo-Nimbus (Cb) clouds that develop locally, beyond 10 km vertically when instability in atmosphere prevails with sufficient levels of temperature and humidity. Both of these are having similar characteristics except that in case of hailstorm, solid precipitation occurs. They occur mostly in summer season (March to May) and less in other seasons. Hail is a rare phenomenon than thunderstorms. Though hail can occur in any part of the world, it is more prevalent in temperate areas. The hail size is proportional to the height of the cloud. Both the thunderstorm and hailstorm operate under similar characteristic phenomenon. They are both associated with lightning, thunder (sound produced by the violent expansion of the air, which can usually be heard up to 8–16 km away), thunderbolt (cloud to ground lightning), heavy spell of rainfall in short duration, and gusty/squally winds of the order of up to 150 kmph. It generally happens in groups, series, and lasts for an hour or two.

6.2.5.1 Effect of Thunder and Hail on Agriculture

Thunder can cause a considerable damage to agricultural crop production if it is associated with high winds, lightening, and heavy rainfall. Due to high winds, there is a chance of uprooting of plants. Thunderbolts can be extremely destructive in terms of fire hazards, and heavy rainfall not only leads to loss of soil nutrients and erosion but at times can be a source of floods as well.

The damage incurred with hails is determined by its characteristics that include the size and number of hailstones that fall per unit area and strength of winds during a high fall. The relationship between intensity and damage depends greatly on the target such as crop, livestock, or property. Certain crops of tender leaf are damaged even by the slight hail; however, maize is not damaged unless the hailstones are of 2 cm or more size. A specific type of hailstorm may not cause much damage during the vegetative phase, but it could be destructive during the flowering and fruit development stages.

6.2.5.2 Management and Mitigation

It includes all the strategies done to prevent the losses of hail in advance of hailstorms using hail-forecasting techniques such as RADAR and probability mapping. The hail control mechanisms prevent the growth and fallout of hailstones in which it is necessary to detect the hail center in a cloud at the right time (using a radio-locator) and then to inject crystallizing reactants. The injection of reactant into clouds is carried out in two ways: (1) bombarding clouds by artillery projectiles containing the crystallizing substance; (2) using counter-hail rockets. Silver iodide and lead iodide are used as crystallizing reactants, which promote cloud growth rather than hailstone development. There are other protection measures such as hail nets and shelter belts.

6.2.6 Drought

Drought is considered by many to be the most complex but least understood of all natural hazards, affecting more people than any other hazard. Drought differs from other natural hazards (e.g., floods, tropical cyclones, and earthquakes) in several ways. First, since the effects of drought often accumulate slowly over a considerable period of time and may linger for years after the termination of the event, the onset and end of drought is difficult to determine. Because of this, drought is often referred to as a creeping phenomenon. Drought is the consequence of a natural reduction in the amount of precipitation received over an extended period of time, usually a season or more in length, although other climatic factors (e.g., high temperatures, high winds, and low relative humidity) are often associated with it in many regions of the world and can significantly aggravate the severity of the event. Drought is also related to the timing (i.e., principal season of occurrence, delays in the start of the rainy season, occurrence of rains in relation to principal crop growth stages) and the effectiveness of the rains (i.e., rainfall intensity, number of rainfall events). Agricultural drought is the impact of meteorological/hydrological drought on crop yield. The agricultural factors that govern the drought impact are the type of crop, stage of plant development, and method of cultivation. Irrigation and anthropogenic factors associated with land use practices, notably deforestation and overgrazing that tend to modify the surface reflectivity (Albedo), surface roughness, and moisture convergence, are the other factors that are also responsible for drought. These have an effect on the feedback on moisture recycling mechanisms, resulting in reduced evaporation and hence accessible atmospheric moisture required for cloud formation and precipitation.

6.2.6.1 Management and Mitigation

There are two distinct phases in which the application of weather and climate knowledge can reduce the impact of drought on communities. The first phase is long-term planning, in which strategies can be devised and precautions taken to reduce impact. The second phase is the action to be taken to reduce the adverse effects during the onset of the event. Since technological inputs quickly reach an optimum level, more emphasis should be placed on drought management policies, especially in dry land farming areas. Agricultural planning and practices need to be worked out with consideration to the overall water requirements within an individual agroclimatic zone. Crops that need a short duration to mature and require relatively little water need to be encouraged in drought prone areas. Irrigation, through canals and groundwater resources, need to be monitored with optimum utilization, avoiding soil salinity and excessive evaporation loss (Das, 1999). By adopting varieties that are drought-resistant at different growth stages, the effects of drought can be avoided or minimized. In high-rainfall areas where there are a series of wet and dry spells, rainfall can be harvested in either farm ponds or in village tanks and can be recycled as lifesaving irrigation during a prolonged dry spell. The remaining water can also be used to provide irrigation for a second crop with a lower water requirement such as chickpea.

6.2.7 Dust Storm

These are the most dangerous winds that blow with the speeds above 10 m/s, bringing plenty of dust and destroying the surface soil layer. These winds, called dust storms, are a phenomenon characterized by the transportation of large volumes of dust and sand by high winds. During dust storms, visibility is impaired significantly. Their occurrence has both natural and human reasons. Agricultural management, not ideally suited to the climatic zone, can cause dust storms. The drying of the upper soil horizon, absence of crop cover, low relative air humidity (below 50%), and the absence of snow cover in winter increases the probability of dust storms. Dust storms may also happen in the steppe zone at the end of summer when the soil dries up and the fields begin to be ploughed after harvesting of the early spring cereals. Winter dust storms are rare phenomena.

6.2.7.1 Effect of Dust Storms on Agriculture

Dust storms are often associated with droughts. Such a combination causes major losses to a national economy. Wind destruction or deflation of soils takes place; crops are ruined as a result of seeds and poor developed seedling being blown away; fields, roads, canals, water reservoirs, irrigation systems are covered by sand and dust; all kinds of infrastructure and communication networks are adversely affected. Wind erosion may remove soils to long distances.

6.2.7.2 Management and Mitigation

Field afforestation is the main measure to protect the soil from dust storms. Improving soil resistance to erosion can be achieved by careful selection of cultivation methods: applying mineral and organic fertilizers, sowing grass, and spraying various substances that enhance soil structure. It is also important to reduce the areas where dust can gather, especially in tracts in areas characterized by erosion. One major protection strategy is to establish well-developed plant cover before the dust storms period.

6.2.8 Heat Wave

It is the persistent period of abnormally high surface temperature comparative to that of normal expected temperature. The time span ranges from few days to few weeks and drops a serious impact on agriculture and allied sector equally. Heat waves are the direct consequence of climate change and it is associated with either low humidity leading to drought or high humidity leading to heat-related stress such as wilting and heatstroke. According to the World Meteorological Organization (WMO), heat wave is defined as "Five or more consecutive days during which the daily maximum temperature surpasses the average maximum temperature by 5°C (9°F) or more above the normal temperature." Another definition as per the WMO is "A marked unusual hot weather (maximum, minimum and daily average) over a region persisting at least two consecutive days during the hot period of the year based on local climatological conditions, with thermal conditions recorded above given thresholds."

Technically, a heat wave is declared when the maximum temperature deviates from normal by +4 °C to +5 °C or more for the regions where the normal maximum temperature is more than 40 °C and the maximum temperature deviates from normal by +5 °C to +6 °C for regions where the normal maximum temperature is 40 °C or less. In other words, heat wave is declared only when the maximum temperature of a station reaches at least 400 °C for plains and at least 300 °C for hilly regions. When actual maximum temperature remains 45 °C or more irrespective of normal maximum temperature, heat wave is declared.

6.2.8.1 Effect of Heat Wave on Agriculture

The impact of a heat waves on agriculture can be direct or indirect. Direct impact includes physical damage to crops, animals, and trees referred to as heat injury caused by the extreme hydro-meteorological events such as hot dry desiccating winds. Heat injury in plants includes scalding and scorching of leaves and stems, sunburn on fruits and stems, leaf drop, rapid leaf death, and reduction in growth. Indirect damage leads to loss of potential production due to disturbed flow of goods and services, decreased or minimal production abilities, hence increased costs of production. Other impacts of heat waves include restricted crop growth and development at different levels such as soil moisture uptake, root and shoot growth, photosynthesis, respiration, plant water uptake, and final yield. Heat waves compete for soil moisture by accelerating evaporation, leaving negligible or almost no moisture in the soil for uptake by plants. Heat waves also cause an overall environmental degradation, which is a major factor contributing to the vulnerability of agriculture, forestry, and rangelands to heat waves.

Other indirect effects of heat waves include poor people working on the agricultural lands. This extreme heat knocks down the agricultural poor and weak people health wise.

6.2.8.2 Management

- **Micro irrigation**: To mitigate the heat waves and to make efficient use of available water to the plants, Drip and Sprinkler irrigation is being followed. This type of irrigation uses 30–50% less water than the standard conventional watering methods. Drip irrigation method helps in supplying water to the root zone of the plants. As a result, water is not wasted on soil. It also reduces the chance of evaporation and run off.
- **PMFBY**: Horticulture farmers can enroll themselves under PMFBY (Pradhan Mantri Fasal Bima Yojana) scheme in which they can avail the insured amount in case of crop loss due to various factors such as high temperature and drought.
- Other moisture conservation techniques followed for horticulture crops are as follows:

1. **Mulching sheet**: The moisture barrier property of plastic film does not allow the soil moisture to escape. Evaporated water is brought back to the surface of the soil by condensation under the plastic film that falls back as droplets, thus restoring the lost moisture. This moisture can be preserved for several days and hence increases the period between two irrigations.
2. **Coir pith**: It is recommended that 5 tons of composted coir pith per hectare of land irrespective of the crop helps in retaining the moisture of soil. It is advised that composted coir pith should be applied basally before take up the sowing. For applying to established trees such as coconut, mango, banana, and other fruit-bearing trees, a minimum of 5 kg composted coir pith is required. It improves the water-holding capacity (more than five times its dry weight), contributing to increased soil moisture.

6.2.9 Cold Wave

Cold wave/frost is a localized seasonal phenomenon prevalent in the country except in Southern India. Frost damage to crops occurs when moisture within the plant is frozen, gets crystallized, and expands. This causes cells to rupture and fluid to leak out, thus the watery appearance of plant tissue or seed after a damaging frost. Different parts of plant, different stages of development of plant, and different types of plants can have varying levels of these "antifreeze" compounds that result in a range of susceptibility to frost.

6.2.9.1 Effect of Cold Wave on Agriculture

The extent of damage caused by cold wave depends on temperature, length of exposure, humidity levels, and the speed at which freezing temperature is reached. It is difficult to predict a definite temperature level up to which crops can tolerate cold wave/frost because many other factors also affect it. Frost and cold waves greatly impact pulse crops. During flowering stage, these crops are likely to be adversely affected at a temperature of −2 to −3 °C. Those in podding stage are a bit more tolerant but are likely to be damaged at a temperature of −3 to −4 °C. As pulses often mature from bottom of the plant to the top, frost injury may be much greater on plant tops.

6.2.9.2 Mitigation

Farmers must provide light irrigation as needed, immediately prune damaged tips of branches or shoots, burn leave/waste material in the orchard to create smoke, and manage rejuvenation of damaged crops through pruning of dead material and application of extra doses of fertilizer through foliar sprays. The measures to be taken against the growth stage of plant are mentioned in Table 6.1.

Table 6.1 Mitigation measures against cold wave for plants

S. no.	Stage of plant growth	Measures to be taken
01	Seeding or nursery stage	Change of microclimate by smoking around the field, especially during night
02	Vegetative stage	Irrigating the field and smoking around the field during night
03	Reproductive stage	
04	Harvesting stage	Harvest the crop at physiological maturity age

Source: https://agricoop.nic.in/sites/default/files/draftdisplan7813.pdf

6.2.10 Low Temperature, Frost, and Snow

Low temperatures during winter mainly cause the injury in plants, referred to as cold injury. Cold injury can be considered when minimum temperature of a day is below −10 °C and severe when it goes −20 °C for longer periods. It often happens in December/January. Low soil temperatures at the depth of plant roots can cause frost injury. Such decreases in soil temperature occur during strong frosts, in the absence of snow cover, and during deep freezing of the soil. It destroys the horticulture produce, hampers germination of seeds, causes chilling injury, and importantly reduces the produce by reducing the season. The damage and destruction is mainly due to extreme cold weather in winter, which results in great economic losses – reduced grain harvests and great expenditure on resowing of destroyed winter crops.

6.2.10.1 Management and Mitigation

The main measures to protect crops from cold injury include the introduction of frost-resistant cultivars and a key factor in protection from cold injury is consistent air temperatures and snow cover throughout the winter.

6.3 Air Pollution

Air pollution refers to the release of pollutants into the air that are detrimental to human health and planet as a whole. Air pollution is the introduction of chemicals, particulates, or biological materials into the atmosphere that cause discomfort, disease, death to humans, damage other living organisms such as food crops, or damage the natural environment or built environment. "Most air pollution comes from energy use and production, Air pollution in the form of carbon dioxide and methane raises the earth's temperature." The problems of air pollution in India have increased since last few decades due to increased population, industrialization, and urbanization. Thermal power plants and transport sector is the biggest contributor of air pollution in the country. The prime air pollutants have been broadly classified as outdoor and indoor pollutants.

Outdoor pollutants include remains of fossil fuel, carbon particles, and metallic particles in the atmosphere from industrial and automobile emissions, toxic gases (i.e., nitrogen dioxide, carbon monoxide, sulphur dioxide, etc.), ozone, tobacco smoke, and so on. On the other hand, indoor pollutants include toxic gases from kitchen fuels and building materials (i.e., asbestos, lead, etc.).

Delhi has been found to have the highest concentration of "respiratory suspended particulate matter" (RSPM) in the air. Particulate matter of small size has been found to be responsible for hypertension in individuals.

6.3.1 Effect of Air Pollution on Agriculture

Agricultural crops can be damaged when exposed to high concentrations of various pollutants. The damage ranges from visible markings on the foliage to reduced growth and yield, to premature death of the plant. The development and severity of the damage also depends on the length of exposure to the pollutant, the plant species and its stage of development, as well as the environmental factors conducive to build up of the pollutant and to the preconditioning of the plant, which make it either susceptible or resistant to injury.

6.3.2 Main Pollutants

6.3.2.1 Oxidants

Ozone is the main pollutant in the oxidant smog complex. Its effect on plants was first observed in the Los Angeles area in 1944. Ozone symptoms characteristically occur on the upper surface of affected leaves and appear as a flecking, bronzing, or bleaching of the leaf tissues. This reduces yield and causes crop loss even without the sign of pollutant stress.

6.3.2.2 Sulphur Dioxide

The major sources of sulphur dioxide are coal burning operations, especially those providing electric power and space heating. Sulphur can also be emitted from the burning of petroleum and the smelting of sulphur containing ores.

Sulphur dioxide enters the leaves mainly through the stomata and the resultant injury can be acute or chronic. Acute injury is caused by absorption of high concentrations of sulphur dioxide in relatively short time. Chronic injury is caused by long-term absorption of sulphur dioxide at sublethal concentrations.

6.3.2.3 Fluorides

Fluorides are discharged into the atmosphere as a result of the combustion of coal; the production of brick, tile, enamel frit, ceramics, and glass; the manufacture of aluminum and steel; and the production of hydrofluoric acid, phosphate chemicals, and fertilizers. Fluorides absorbed by leaves are carried to the margins of broad leaves and the tips of monocotyledonous leaves. Little injury takes place at the site of absorption, but the margins or the tip of the leaves build up injurious concentrations.

6.3.2.4 Ammonia

Ammonia is released into the environment mainly by agricultural practices and pharmaceutical companies. Ammonia injury is severe to vegetation in the immediate vicinity. Complete system expression on affected vegetation usually takes several days to develop and appears as irregular, bleached, bifacial, necrotic lesions.

6.3.2.5 Particulate Matter

Particulate matter such as cement dust, magnesium-lime dust, and carbon soot deposited on vegetation can inhibit the normal respiration and photosynthesis mechanisms within the leaf. Cement dust may cause chlorosis and death of leaf tissue by the combination of a thick crust and alkaline toxicity produced in wet weather. The dust coating may also affect the normal action of pesticides and other agricultural chemicals applied as sprays to foliage. In addition, accumulation of alkaline dusts in the soil can increase soil pH to levels adverse to crop growth.

In India, in 2014, it was reported that air pollution by black carbon and ground level ozone had reduced crop yields in the most affected areas by almost half in 2011 when compared to 1980 levels.

6.4 Role of Meteorology in Air Pollution

Meteorology plays a vital role in determining the concentrations of the pollutants. The fundamental way in which it interacts is the dispersion or the dilution of the pollutants. Meteorology governs the pollutants completely once they are emitted into the atmosphere. Air movement influences the fate of air pollutants, so any study of local weather patterns (meteorology) is beneficial. If the air is calm and pollutants cannot disperse, then the concentration of these pollutants will build up. On the other hand, when strong, turbulent winds blow, pollutants disperse quickly, resulting in lower pollutant concentrations.

Meteorological data help:

- Identify the source of pollutants;
- Predict air pollutions events such as inversions and high-pollutant concentration days; and
- Simulate and predict air quality using computer models.

When studying air quality, it is important to measure the following factors as they can help us understand the chemical reactions that occur in the atmosphere:

- Wind speed and direction
- Temperature
- Humidity
- Rainfall
- Solar radiation

6.4.1 Wind Speed and Direction

Wind data records can determine the general direction and the area of emissions. Identifying the sources means planning to reduce the impacts on air quality. An instrument called anemometer measures wind speed, while a wind vane informs about its direction. Winds are usually summarized by classes of speeds and directions. Speed determines the dilution of the pollutants in the atmosphere to some extent, while direction determines the transportation. A sonic anemometer operates on the principle that the speed of wind affects the time it takes far sound to travel from one point to another. Sound travelling with the wind will take less time than sound travelling into the wind. By measuring sound wave speeds in two different directions at the same time, sonic anemometers can measure both wind speed and direction.

6.4.2 Temperature

Measuring temperature supports air quality assessment, air quality modeling, and forecasting activities. Temperature and sunlight play an important role in the chemical reactions that occur in the atmosphere to form photochemical smog. The most common way of measuring temperature is to use a material with a resistance that changes with temperature such as platinum wire. A sensor measures this change and converts it into a temperature reading. The vertical temperature variation plays an important role in air pollution. The increase in temperature with height in the low level is of great concern because, depending on the level at which it occurs with regard to the stack height, such inversion can act as a lid and not allow dispersion to take place, resulting in enormously high ground level concentrations.

The spatial variation of surface temperature also plays a role in air pollution studies. The surface temperature in the cities is usually higher compared to nearby rural areas due to urbanization known as "urban heat island" (UHI). The UHI influences the dispersion of pollutants. Stability is another important parameter that governs the dispersal of pollutants.

Highly unstable conditions result in thorough mixing and dilution and a consequent reduction in the ground level concentration. Stable conditions, on the other hand, are characterized by low mixing and cannot disperse the pollutants resulting in the buildup of pollutants.

6.4.3 Humidity

Like temperature and sunlight, water vapor plays an important role in many thermal and photochemical reactions in the atmosphere. As water molecules are small and highly polar, they can bind strongly to many substances. If attached to particles suspended in the air, they can significantly increase the amount of light scattered by the particles (measuring visibility). If water molecules attach to corrosive gases, such as sulphur dioxide, the gas will dissolve in the water and form an acid solution that can damage health and property.

Water vapor content of air is reported as a percentage of the saturation vapor pressure of water at a given temperature. This is the relative humidity. The amount of water vapor in the atmosphere is highly variable, and it depends on geographic location, how close water bodies are, wind direction, and air temperature. Relative humidity is generally higher during summer when temperature and rainfall are also at the highest.

Measuring humidity uses the absorption properties of a polymer film. The film either absorbs or loses water vapor as the relative humidity of the ambient air changes. A sensor measures these changes and coverts them into a humidity reading.

6.4.4 Rainfall

Rain has a "scavenging" effect because it washes particulate matter out of the atmosphere and dissolves gaseous pollutants, removing particles and so improving visibility. If rain dissolves gaseous pollutants such as sulphur dioxide, it can form acid rain, resulting in potential damage to materials or vegetation. A common method to measure rainfall is to use a tipping bucket rain gauge. The instrument calculates the quantity and intensity of rainfall based on the area of the funnel and the rate of bucket movements.

6.4.5 Solar Radiation

It is important to monitor solar radiation for use in modeling photochemical smog events, as the intensity of sunlight has an important influence on the rate of the chemical reactions that produce the smog. The cloudiness of the sky, time of the day, and geographical location all affect sunlight intensity. An instrument called pyranometer measures solar radiation from the output of a type of silicon cell sensor.

Bibliography

Agrawal M (2005) Effects of air pollution on agriculture: an issue of national concern. Natl Acad Sci Lett 28(3/4):93–106

Ahmad L, Qayoom S (2019) Precipitation and rainy day trends in Kashmir Valley, Jammu & Kashmir, India. Indian J Ecol 46(1):83–87

Ahmad L, Kanth RH, Parvaze S, Mahdi SS (2017) Experimental agrometeorology: a practical manual. Springer, Cham, pp 83–87

Arciszewska C, McClatchey J (2001) The importance of meteorological data for modelling air pollution using ADMS-Urban. Meteorol Appl 8(3):345–350

Bal SK, Saha S, Fand BB, Singh NP, Rane J, Minhas PS (2014) Hailstorms: causes, damage and post-hail management in agriculture. Technical bulletin 5:44

Bhaskar BV, Mehta VM (2010) Atmospheric particulate pollutants and their relationship with meteorology in Ahmedabad. Aerosol Air Qual Res 10(4):301–315

Bony S, Stevens B, Carlson D (2017) Understanding clouds to anticipate future climate. WMO Bull 66:8–11

Brenner R, Brownell W, Walke J (2010) Air pollution standards for stationary sources – next moves. Environ Rep News Analysis 40:10012

Carter WN (2008) Disaster management: a disaster manager's handbook. Asian Development Bank, Manila, Philippines

Choudhary MP and Garg V (2013) Causes, consequences and control of air pollution. In All India Seminar on Methodologies for Air Pollution Control, held at MNIT

cseindia (n.d.) http://www.cseindia.org/content/workshop-global-burden-disease-air-pollution-amongst-top-killers-india

Das HP, Adamenko TI, Anaman KA, Gommes RG, Johnson G (2003) Agrometeorology related to extreme events. WMO, Geneva

District Disaster Management Agency, Vellore (2018) Prevention and management of heat waves. https://cdn.s3waas.gov.in/s31651cf0d2f737d7adeab84d339dbabd3/uploads/2018/06/2018062917.pdf

Ebenezer GO (2019) The role of meteorology in atmospheric processes and air pollution studies

eschooltoday (n.d.) http://eschooltoday.com/pollution/air-pollution

Gabel K (2011) Monitoring pollution in our communities: the Clean Air Coalition of Western New York

Gbeckor-Kove N (1995) Main climate events affecting agricultural production. 4 Eme symposium international Sur1. In: Assurance Agricole. WMO, Switzerland

Gommes R and Nègre T (1992) The role of agrometeorology in the alleviation of natural disasters. FAO Agrometeorology Series Working Paper (FAO). no. 2

Gommes R, Tychon B and Tonnard V (1997) Prévision agrométéorologique des rendements: quelques moyens et méthodes utilisés par la FAO dans un contexte de sécurité alimentaire

Guttikunda S (2010) Role of meteorology on urban air pollution dispersion: a 20 yr analysis for Delhi, India. SIM-air Working Paper Series. 31

Hagman G, Beer H, Bendz M and Wijkman A (1984) Prevention better than cure. Report on human and environmental disasters in the Third World. 2

Mallet MD (2021) Meteorological normalisation of PM10 using machine learning reveals distinct increases of nearby source emissions in the Australian mining town of Moranbah. Atmos Pollut Res 12(1):23–35

Muzamil SAHBS, Zainun NY, Ajman NN, Sulaiman N, Khahro SH, Rohani MM, Mohd SMB, Ahmad H (2022) Proposed framework for the flood disaster management cycle in Malaysia. Sustainability 14(7):4088

National Institute of Environmental Health Sciences (n.d.) http://www.niehs.nih.gov/health/topics/agents/airpollution

NRDC (n.d.) http://www.nrdc.org/air/

omafra.gov.on (n.d.) http://www.omafra.gov.on.ca (Air pollution on Agricultural crops, June 2003)

Pollution A (1998) MN Rao & HVN Rao, Tata McGraw-Hill, 2007. retrieved at https://www.scribd.com/document/337844888/M-N-Rao-H-V-N-Rao-Air-Pollution-pdf

Rafferty JP (2018) Heat wave. Encyclopedia Britannica

Srikanth D (n.d.) Role of IMD in meteorological disaster management. Ministry of Earth Sciences, India. India Meteorological Department. http://www.imdhyderabad.gov.in/

Tannehill IR (1947) Drought, its causes and effects 64, 1, 83. LWW Philadelphia

The Guardian (2014) India air pollution 'cutting crop yields by half' *The Guardian*, 3 November 2014

urbanemissions.blogspot.in (n.d.) http://urbanemissions.blogspot.in/2014/01/delhi-ranks-ist-among-world-cities-with.html

WHO (n.d.) http://www.who.int/topics/air_pollution

Wikipedia (n.d.) http://en.wikipedia.org/wiki/Air_pollution

Wu M, Chen Y, Wang H, Sun G (2015) Characteristics of meteorological disasters and their impacts on the agricultural ecosystems in the northwest of China: a case study in Xinjiang. Geoenviron Disasters 2(1):1–10

Weather Forecasting

7

Abstract

Weather forecasting is the scientific application that helps to predict the atmospheric conditions for a specific time and location. Weather forecasting is an effective tool in the modern agricultural practices. Climate change, however, influences the weather and makes it difficult to predict. Climate change results in large-scale shifts in weather patterns, which are primarily due to global warming caused by greenhouse gas emissions by human. This chapter is a compact of all the concepts of weather forecasting, global warming, and climate change. In this chapter, remote sensing as an aid of forecast and atmospheric chemistry has also been discussed in detail.

Keywords

Weather forecasting · Global warming · Greenhouse gases · Climate variability · Climate change · Atmospheric chemistry · Remote sensing · El Nino · La Nina · ENSO

Weather forecasting is the prediction of the weather through application of the principles of physics, supplemented by a variety of statistical and empirical techniques. In addition to predictions of atmospheric phenomena themselves, weather forecasting includes predictions of changes on the earth's surface caused by atmospheric conditions, e.g., snow and ice cover, storm tides, and floods. According to science, forecasting means the process of estimation of the value of some variable at some future time. One of the primary functions of the national weather services is forecast of weather parameters such as rainfall, temperature, wind, and humidity over a region averaged over a particular time period, e.g., forecast of daily rainfall (rainfall averaged over a day).

L. Ahmad et al., *Climate Change and Agrometeorology*,
https://doi.org/10.1007/978-981-99-4863-5_7

Weather forecasts are made by collecting as much data as possible about the current state of the atmosphere (particularly temperature, humidity, and wind) and using understanding of atmospheric processes (through meteorology) to determine how the atmosphere evolves in the future. However, the chaotic nature of the atmosphere and incomplete understanding of the processes mean that forecasts become less accurate as the range of the forecast increases. In some ways, weather forecasting is still carried out in the same way as it was done by the humans earlier, such as by making observations and predicting changes. No doubt, the modern tools used to measure temperature, pressure, wind, and humidity in the twenty-first century is certainly better. Yet, even the most sophisticated numerically calculated forecast made on a supercomputer requires a set of measurements of the condition of the atmosphere, i.e., an initial picture of temperature, wind, and other basic elements, which is somewhat comparable to that formed by our forebears when they looked out of their cave dwellings. The primeval approach entailed insights based on the accumulated experience of the perceptive observer, while the modern technique consists of solving equations. Although seemingly quite different, there are underlying similarities between both practices. In each case the forecaster asks "What is?" in the sense of "What kind of weather prevails today?" and then seeks to determine how it will change in order to extrapolate what it will be.

The basis for weather prediction started with the theories of the ancient Greek philosophers and continued with Renaissance scientists, the scientific revolution of the seventeenth and eighteenth centuries and the theoretical models of twentieth- and twenty-first-century atmospheric scientists and meteorologists. Likewise, it describes the development of the "synoptic" idea – that of characterizing the weather over a large region at exactly the same time in order to organize information about prevailing conditions. In synoptic meteorology, simultaneous observations for a specific time are plotted on a map for a broad area whereby a general view of the weather in that region is gained. (The term synoptic is derived from the Greek word, meaning "general or comprehensive view.") The so-called synoptic weather map came to be the principal tool of nineteenth-century meteorologists and continues to be used today in weather stations and on television weather reports around the world.

Nowadays, the widespread adoption of numerical weather prediction models brought a whole new group of players, i.e., computer specialists and experts in numerical processing and statistics so as to work with atmospheric scientists and meteorologists.

7.1 Principles and Methodology of Weather Forecasting

7.1.1 Short-Range Forecasting

Short-range weather forecasting provides information on the expected weather with forecast projection times ranging from a few hours to 2 or 3 days for both particular locations and areas covering a few million square kilometers. Almost all currently used short-range forecasting techniques involve dynamic prediction models based on an application of compressible fluid mechanics equations to the atmosphere. The models provide information necessary not only for the weather prediction but also for many other applications (aviation, ship routing, etc.). As a rule, the models are running within computerized real-time forecasting systems, which involve automated collection, processing, checking, and numerical analysis of observations necessary for weather forecasting. Short-range forecasts can be carried out by a number of ways, some of which are mentioned as follows.

7.1.1.1 Point Forecasts

In point forecasts, we combine available weather forecast data with the TESLA Model to generate highly accurate and reliable point forecasts for horizons extending to the limit of the available weather forecast. Here, we monitor the performance of our point forecasts against real-time observations.

7.1.1.2 Comparable Days

In comparable days, the developed algorithms are used to scan the available histories of observed weather data to identify as many as 5 days in the past that are meteorologically comparable to the forecast day. Our system will then display the observed demand data from those days alongside our point forecasts. While demand systems do change over time, the Comparable Days tool can lend valuable insight into how a system is capable of responding to specific weather scenarios.

7.1.2 Medium-Range Forecast

A forecast for a period extending from about 3–7 days in advance; there are no absolute limits to the period embraced by the definition.

7.1.3 Long-Range Forecast

As per the World Meteorological Organization (WMO) definition, long-range forecast is defined as the forecast from 30 days up to one season's description of averaged weather parameters. The monthly and seasonal forecast comes under long-range forecast.

7.1.3.1 What Are the Different Methods Used for Long-Range Forecasting?

In general, there are three approaches used for long range forecasting and these are (1) statistical method, (2) numerical weather prediction or dynamical method, and (3) dynamical-cum-statistical method. From the beginning, the main approach toward the long-range prediction has been based on statistical methods. The India Meteorological Department's (IMD) operational forecasts for the monsoon rainfall are based on this technique. The statistical method involves identifying predictive signals (predictors, i.e., atmospheric variables) that have a significant and stable historical relationship with the predict (atmosphere) and predicting the value of predict and at future time. For this purpose, it is assumed that the observed predict and predictor relationship persists in the future also and that the predictor values corresponding to the future predict and value to be forecasted are known.

An alternate approach toward the prediction of Indian summer monsoon rainfall (ISMR) is based on numerical models or the so-called General Circulation Models (GCMs) for the simulation of atmospheric and oceanic conditions. Though the numerical prediction models have the potential for providing predictions over smaller spatial and temporal scales as per the user demands, they have not so far shown required skill for simulating the salient features of the mean monsoon rainfall and its interannual variability. For improved rainfall simulations, GCMs should be able to account for local sub-grid features and sub-seasonal variability of the climatic fields.

The third approach of dynamical-cum-statistical method is based on the fact that GCMs have better skill in simulating large-scale atmospheric circulation features and that there exists a semiempirical relation between rainfall over a region and prevailing large-scale circulation features over both global and regional scales. Therefore, it is possible to derive recalibration equations between the rainfall and circulation features simulated by GCMs, and assuming that these relationships will hold good in the future, regional rainfall can be predicted. The dynamical-cum-statistical method is a recent development in the long-range prediction.

The India Meteorological Department is solely responsible for issuing operational long-range forecast for India. The forecasts are prepared at the National Climate Center of IMD located at Pune. At present, empirical (statistical) methods are used for the preparation of operational long-range forecasts.

The IMD issues operational long-range forecast for the rainfall during SW Monsoon season (June–September). These forecasts are issued in two stages. The first stage forecast is issued in mid-April and consists of quantitative forecast for the season (June–September) rainfall over India as a whole. The second stage forecasts issued by the end of June consist of update for the forecast issued in April, a forecast for July rainfall over the country as a whole and forecasts for seasonal rainfall over broad rainfall homogeneous regions of India.

The IMD also prepares forecasts for winter (January–March) precipitation (issued in the end of December) over Northwest India and Northeast Monsoon (October–December) rainfall over Southern Peninsula (issued in October). However, these forecasts are issued only to the government and they are broadly classified into the following types, as mentioned in Table 7.1.

7.2 Southern Oscillation: El Nino, La Nina, and ENSO

Southern oscillation or "SO" is a "see-saw" in the surface air pressure between eastern and western tropical Pacific. It is characterized by simultaneously opposite sea level pressure anomalies at Tahiti, in the eastern tropical Pacific and Darwin, on the northwest coast of Australia. The SO was discovered by Sir Gilbert Walker in the

Table 7.1 Forecasting ranges

Range	Time	Methods	Phenomenon	Utility
Nowcasting	6 h	Radars	Hailstorms, squalls, with high accuracy	Severe weather warnings (500 m)
Short range	2–3 days	Nested atmospheric models	Synoptic scale weather system	Conventional forecasting resolution (3–25 km)
Medium range	7–10 days	Global atmospheric models	Synoptic scale weather system	Conventional forecasting resolution (25–50 km)
Extended range	10–30 days	Coupled atmospheric models	Persistent systems • Blocking highs • MJO • ITCZ	Drought and heat/cold waves

early 1920s. Later, the three-dimensional east–west circulation related to the SO was discovered and named the Walker Circulation. The SO has periodicity of about 2–5 years. A most common index of SO is computed as the difference between standardized sea level pressure anomalies at Tahiti and Darwin (Tahiti–Darwin).

- El Nino and La Nina are oceanic manifestation of opposite phases of SO, which is an atmospheric phenomenon. El Nino is characterized by the warming of the sea surface temperatures (SSTs) in the central and eastern equatorial Pacific Ocean, beginning at about Christmas time (hence the name "El Nino," which is a reference to the Christ child). This is called the warm phase of the SO. The cold phase of the SO, called "La Nina," is characterized by high pressure in the eastern equatorial Pacific, low in the west, and by anomalously cold SST in the central and eastern Pacific.
- ENSO (El Nino Southern Oscillation) is an acronym designed to stress the fact that the El Nino and SO are components of the same global ocean–atmosphere coupled phenomena.

7.2.1 What is the Relationship Between Monsoon and ENSO?

Both monsoon and ENSO are ocean–atmosphere couple phenomena. There is a general inverse relationship between monsoon and ENSO. The warm phase of ENSO is generally associated with weaker than normal monsoon and vice versa. During the period 1885–2007, there were 36 years of warm ENSO (El Nino) and 25 years of cold ENSO (La Nina). During the 15 of the 35 El Nina years (42%), Indian summer monsoon rainfall (ISMR) was below normal, and during the 9 of the 25 La Nina years (36%), ISMR was above normal. This shows that there is no one-to-one correspondence between ENSO and ISMR.

7.3 Remote Sensing

Remote sensing is defined as "the science and art of obtaining information about an object, area, or phenomenon through the analysis of data acquired by a device that is not in contact with the object, area or phenomenon under investigation."

The general term originated in the 1960s, at a similar time to the launch of the first meteorological satellite, the Television InfraRed Observation Satellite (TIROS-1). Remote sensing is the process of detecting and monitoring the physical characteristics of an area by measuring its reflected and emitted radiation at a distance (typically from satellite or aircraft). Special cameras collect remotely sensed images, which help researchers "sense" things about the earth. Some examples are:

- Cameras on satellites and airplanes take images of large areas on the earth's surface, allowing us to see much more than we can see when standing on the ground.
- Sonar systems on ships can be used to create images of the ocean floor without having to travel to the bottom of the ocean.
- Cameras on satellites can be used to make images of temperature changes in the oceans.

The use is rising in the fields of meteorology and climatology, and it works in tandem with the use of Geographical Information Systems (GIS). Techniques can provide increased spatial coverage when compared to weather station data and the instantaneous observations, and global coverage and improving the quality of remotely sensed information is becoming increasingly useful. Remote sensing offers the ability to work at a number of scales ranging from local to citywide. Regardless of the scale of the study, remote sensing offers the opportunity to provide a consistent and repeatable methodology, suited equally to both quick pilot studies as well as long-term monitoring campaigns. Although the initial cost of remote sensing platforms is high, the ease of data availability to end researchers, combined with the often-extensive temporal and spatial coverage available, offers a marked improvement to traditional fieldwork campaign studies.

Recently, remote sensing has been used for improving understanding of the climate system and its changes. It enables to monitor the earth surface, ocean, and the atmosphere at several spatiotemporal scales, allowing climate system observations, as well as to investigate climate-related processes or long- and short-term phenomena, as for example deforestation or El Niño trends. Moreover, remote sensing is useful to collect information and data in dangerous (e.g., during fire events) or inaccessible areas (e.g., impervious areas). Specific examples of remote sensing uses, also related to climate change adaptation practices, include (1) natural resource management, (2) management of agricultural practices related to land use, land conservation, and soil carbon stock, (3) tactical forest firefighting operations in real-time decision support systems, (4) monitoring of land cover and its changes over different temporal and spatial scales, even after a disaster event, (5) better-informed forest and water management, (6) evaluation of carbon stocks and related dynamics,

(7) simulation of climate system dynamic, and (8) improvement of climate projections and meteorological reanalysis products, widely used for climate change research studies.

Finally, remote sensing can be used for improving warning and preparedness, being therefore also useful in disaster risk management. Geographic Information Systems (GIS) using satellite technology can be used for developing early warning and forecasting systems to reduce and manage climate-related disaster risk (i.e., preparing better prediction of cyclone and flood tracks, drought events, fire occurrence), as well as helping to be prepared for actions. Remote sensing technology can also be useful for post-disaster damage detection based on comparative analysis of before and after disaster images. Remote sensing data and information are also useful for emergency workers.

Some specific uses of remotely sensed images of the earth include:

- Large forest fires can be mapped from space, allowing rangers to see a much larger area than from the ground.
- Tracking clouds to help predict the weather or watching erupting volcanoes, and help watching for dust storms.
- Tracking the growth of a city and changes in farmland or forests over several years or decades.
- Discovery and mapping of the rugged topography of the ocean floor (e.g., huge mountain ranges, deep canyons, and the "magnetic striping" on the ocean floor).

7.3.1 Success and Limiting Factors

Remote sensing techniques, and specifically satellite images, have been already successfully used in a wide range of climate change fields such as for:

1. Investigating global temperature trends, both at the ocean surface and in the atmosphere;
2. Detecting changes in solar radiation affecting global warming;
3. Monitoring aerosols, water vapor concentration, and changes in precipitation regime;
4. Studying the dynamics of snow extension and ice cover;
5. Monitoring sea-level changes and coastal modifications;
6. Monitoring vegetation status and change;
7. Monitoring water resources and impact due to droughts and dry periods;
8. Monitoring fire events and fire emissions;
9. Predicting disaster risk, such as cyclone, floods, and drought; and
10. Guiding decision-making processes on climate change adaptation.

The use of remote sensed data is quickly evolving, both in terms of available techniques and resolution, and other uses relevant for climate change adaption are expected to emerge in the next future.

Some concerns, however, have been posed on the use of remote sensing. Studying and monitoring climate change require long-term time series of observations, while satellite data are often available for short-term period. Furthermore, some uncertainties and distortions of received image frames due to vibrations and turbulence can derive by biases in sensors and retrieval algorithms, so the use of satellite observations in climate change studies requires a clear identification of such limitations. Other possible limitations include: (1) high cost of acquiring aircraft and high-resolution data from drones; (2) limited access to required technologies owing to cost or skill constraints in some cases; and (3) temporal discontinuity of aircraft and satellite data. While the first one can be particularly expensive and hence available for a limited number of surveys, the second and third are collected at fixed intervals depending on the satellite return time.

7.4 Atmospheric Chemistry

Atmosphere is a combination of two Greek words "atmos" meaning vapor and "sphaira" meaning globe, which got evolved in modern Latin as "atmosphaera" and eventually into "atmosphere" in the mid-seventeenth century. A magnanimously vast horizon of the earth's atmosphere is mainly composed of seven layers among which about 90% of total atmospheric mass resides in the troposphere and the bulk of minor trace-gas burden is also found there. Some of the species are liberated from the surface or interior by purely geological sources, gently over long periods or violently in volcanic eruptions (e.g., H_2S, SO_2) and some are generated in thunderstorms (e.g., CO, NO), but many of the substances are present because they or the precursors are generated by living organisms (e.g., CH_4, N_2O).

Air is essentially a homogenous mixture up to an altitude of about 100 km; this region is called homosphere. The two main constituents of air in this region are molecular nitrogen (78% by volume) and molecular oxygen (21% by volume). These are followed by water vapor up to 4% by volume and then along way behind by argon 0.93% by volume and CO_2 0.036% by volume or 260 ppmv but increasing by 1.5 ppmv each year. O_2 and N_2 can absorb the earth's radiations and photodissociate. Thus, atomic species can be produced that include nitrogen atoms and oxygen atoms in both their ground state and O (3p) and first excited electronic state, O (1d). In addition, O_3 is produced and its photodissociation leads to the production of first excited electronic state of molecular oxygen, O_2. Several other species also exist in low quantities in the atmosphere. These include the rare gases, hydrogen, and carbon compounds. The concentrations of N_2, O_2, CO_2, and the rare gases are constant throughout the homosphere ($\leq$85 km). The concentrations of various gases present in the atmosphere along with the symbols are shown in Table 7.2.

In addition to these, there are so many free radicals and ions, as well as other minor constituents containing sulfur, halogens, and metals. Some of these species are present as aerosols. Air is never perfectly clean in the sense of containing only N_2, O_2, CO_2, H_2O, and other inert gases. Even before humans existed, there were many sources of trace constituent species, some of which are now classified as

Table 7.2 Concentrations of various gases of atmosphere

Gas	Symbol	Concentration (ppm)
Nitrogen	N_2	780,840
Oxygen	O_2	209,460
Carbon dioxide	CO_2	416
Helium	He	5.24
Neon	Ne	18.18
Argon	Ar	9340
Krypton	Kr	1.14
Xenon	Xe	0.087
Water vapor	H_2O	Variable
Methane	CH_4	1.4
Hydrogen	H_2	0.5
Carbon monoxide	CO	0.08
Ozone	O_3	0.025
Nitrogen oxide	NO	2×10^{-3}
Ammonia	NH_3	6×10^{-3}
Nitrous oxide	N_2O	0.25
Sulfur dioxide	SO_2	1.00
Nitrogen dioxide	NO_2	4×10^{-2}
Iodine	I_2	0.01

pollutants. Natural air pollution is caused by volcanic eruptions, breaking waves, windblown dust, forest fires and pollens, dust and terpenes from plants. Agricultural practices have been a source of additional pollutant species. "Slash and burn" clearing of lands greatly adds to the fire source. The removal of natural vegetation cover itself increases the rate of dust erosion. Nitrogen used as fertilizer is in part returned to atmosphere as N_2O and alterations in the types and utilization of crops also affect the N_2O budget. Increased population of livestock and rice paddy cultivation implies an increase in CH_4 production.

Combustion of fossil fuel affects the atmosphere in three different ways:

- By the formation of carbon dioxide,
- By the release of substances such as sulfur dioxide and partially oxidized or unburn fuel, and
- By the high temperature fixation of atmospheric N_2 and O_2 to yield oxides of nitrogen.

The carbon dioxide problem arises because in the course of a few hundred years appreciable fractions of earth fuel deposit shave been burnt to release carbon dioxide, thus reversing the chemistry that photosynthesis took milions of time as long to achieve. At present atmospheric CO_2 concentations are by about one part in 300 a year, corresponding to about one-fifth of annual release of CO_2 from combustion of fossil fuels the climatic effects of release in greenhouse gases is greenhouse effect in atmosphere. Coal and petroleum products contain appreciable quantities of sulfur compounds. Combustion then leads to sulfur dioxide release.

Particulate matter (soot), carbon monoxide, and unburnt hydrocarbons are also important minor by-products of the combustion process. Sulfur dioxide SO_2 and NO_x are formed when combustion is supported by air to contribute to problems of acid rain, and oxides of nitrogen also play a special part in 'photochemical air pollution'.

Soot, SO_2, and malodorous compounds emitted from factories contribute to local air pollution. More wide spread dispersal within the troposphere can lead to regional pollution of quite large geographical areas. If the entire atmosphere is affected, then the pollution becomes global in scope.

The London type of air pollution characterized by particulate matter and Sox has been recognized for centuries in the mid-1940s. The effects of a new type of oxidizing air pollution, which caused eye irritation, plant damage, and visibility degradation, became evident in Los Angeles. The oxidants include ozone, nitrogen dioxide, and peroxyacetyl nitrate (PAN) and they are formed photochemically by the action of solar radiations on combinations of NOx and hydrocarbons (HC) in air. Because of the region where it was first observed and because of its photochemical origin, oxidizing pollution of the type described is called Los Angeles smog or photochemical smog.

7.4.1 Layers of Atmosphere

The atmosphere appears to be nothing more than thin air, but scientists have shown that it has five or more layers or "spheres," each with its own characteristics. We live in the lowest layer, called troposphere. This layer is just 7–12 miles (12–18 km) thick, but 75% of atmosphere's gases sink into this layer. As you go higher and higher, the gases get more and more spread out until by the time you are about 600 miles (1000 km) up, the gases are so thin or "rarefied" that the atmosphere fades off into space.

The detailed structure of atmosphere was given by "Sutcliffe" in 1966. The atmosphere is divided into seven layers: exosphere, ionosphere, thermosphere, mesosphere, ozone layer, stratosphere, and troposphere.

7.4.2 Principal Layers of Atmosphere Based on Temperature

On the basis of vertical temperature variation, the atmosphere is divided into different spheres or layers. The earth's atmosphere is divided (called atmospheric stratification) into five main layers. Excluding the exosphere, the earth has four primary layers, which are the troposphere, stratosphere, mesosphere, and thermosphere. From highest to lowest, the five main layers are:

- Exosphere: 700–10,000 km (440–6200 miles)
- Thermosphere: 80–700km (50–440 miles)

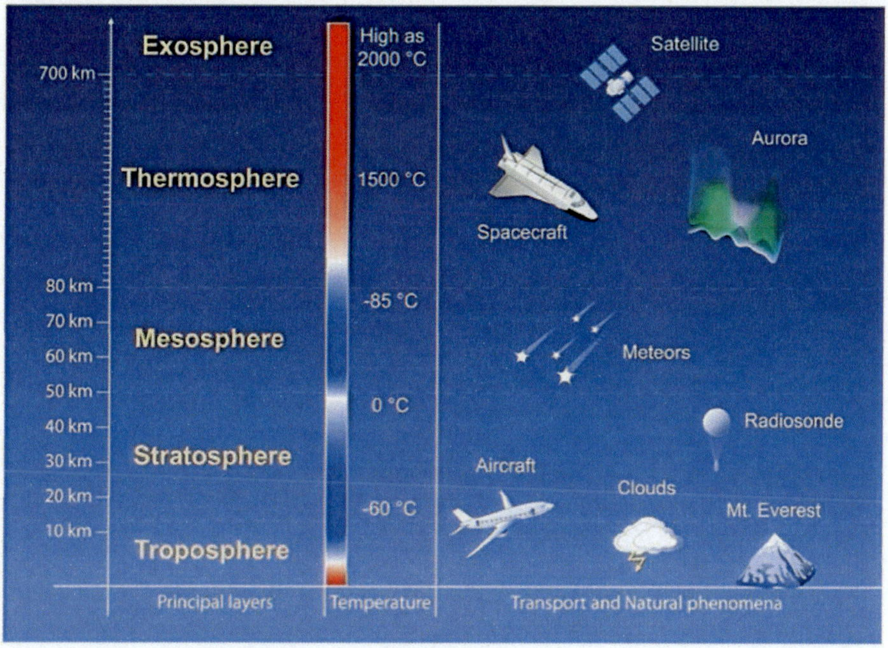

Fig. 7.1 Layers of atmosphere

- Mesosphere: 50–80 km (31–50 miles)
- Stratosphere: 18–50 km (12–31 miles)
- Troposphere: 0–18 km (0–12 miles)

The structure of atmosphere has been considered in a number of ways, also by its physical properties such as temperature, diffusion, and ionization. Figure 7.1 depicts the spheres of the earth's atmosphere and the properties of each sphere has been described in Table 7.3.

7.4.2.1 Troposphere

"Tropo" is a Greek word which means "turning." Turbulent mixing is relevant in this layer. It is the basal part of the atmosphere, which is about 18 km above the earth's surface. This is lowest at the poles, where it is about 7–10 km above the earth's surface. It is highest (about 17–18 km) near the equator. It is the atmospheric layer closes to the planet and contains the largest percentage (around 80%) of the mass of the atmosphere. It contains 70% of the gases and all the weather. For organisms this is the most important zone in the atmosphere.

The important events such as cloud formation, lightening, and thundering all take place in this zone. At the top of the troposphere, there is a shallow layer separating it from the stratosphere known as the "tropopause." The tropopause layer is thin and its height changes according to the latitudes and it is a transitional zone and distinctly characterized by no major movement of air.

Table 7.3 Properties of spheres of the earth's atmosphere

Name of the layer	Height from the earth	Properties
Exosphere	Up to 10,000 km and beyond space	–
Thermosphere	Up to 400 km (D, E, F, and G layers)	Density is less, oxygen and nitrogen present, molecular and atoms of oxygen present, temperature increase with height. Temperature up to 980 °C
Mesopause		
Mesosphere	Around 80 km from the earth	Temperature decreases to a minimum (−90 °C) of around 80 km
Stratopause	Maximum temperature with absorption of UV rays	
Stratosphere (ozonosphere)	50 km from troposphere	Temperature constant up to 20 km, increases up to 50 km
Tropopause	Temperature inversion layer (warm air over cold – 28 ° C)	–
Troposphere	16–18 km at equator and 7–10 km from poles	75% of total gas, water vapor, and aerosols. Decreases in temperature, with height at 6.5 °C/ km. This is up to (−)28 °C

7.4.2.2 Stratosphere

It has been derived from a Latin word "stratum" meaning "blanket." It was first discovered by Leon-Philippe Teisserenc De Bort and Richard Mann in 1900. Temperatures in the stratosphere are highest over the summer pole and lowest over the winter pole. Teisserenc De Bort gave the names troposphere and stratosphere of the lower two layers. It is the second zone of the atmosphere, about 30 km height and 50 km above the earth's surface. This layer is called as "Seat of photochemical reactions."

Due to the presence of ozone gas acting as a layer, stratosphere is also known as ozonosphere, and because of the preponderance of chemical process, this sphere is called as the "chemosphere." The temperature remains almost constant at around 20 km and is characterized as is other mal because the air is thin, clear, cold, and dry near the tropopause, and airlines also cruise in this still air. The altitude of maximum temperature is known as stratopause. The temperature values increase up to 0 °C. This increase in temperature is due to zone formation under the influence of UV component to sunshine. Less convection takes place in the stratosphere because it is warm at the top and cold at the bottom. There is also persistence of circulation patterns and high wind speeds.

The upper boundary of the stratosphere is called the stratopause. A property of the ozone is that it absorbs UV rays. Had there been no layer of the ozone in the atmosphere, the UV rays might have reached the surface of the earth and no life can exist. The temperature of the ozonosphere is high (warm) due to selective absorption of UV radiation by ozone.

7.4.2.3 Mesosphere

"Mesos" in Greek means "middle" because it is in the middle of the five layers. It is about 40 km in height, up to 80 km above the surface. It is too thin in gases to soak up much heats of it, the coldest layer. Here, the temperature again decreases with height, reaching a minimum of about −90 °C at the "mesopause," which is the upper boundary of this layer. Just below the mesopause, the air is so cold that even the very scarce water vapor at this altitude can be sublimated into polar-mesospheric noctilucent clouds. These are the highest clouds in the atmosphere and may be visible to the naked eye if sunlight reflects off them about an hour or two after sunset or a similar length of time before sunrise. The mesosphere is also the layer where most meteors burn up upon atmospheric entrance. It is too high above the earth to be accessible to jet-powered aircraft and balloons and too low to permit orbital spacecraft. The mesosphere is mainly accessed by sounding rockets and rocket-powered aircraft.

7.4.2.4 Thermosphere (Ionosphere)

The thermosphere layer lies beyond the ozonosphere (stratosphere) at a height of about 80 km above the earth's surface and extends up to 400 km. "Thermos" is a Greek word meaning "warm heat." This layer can be heated to above 1000 °C by the sun. This temperature increase is caused by the absorption of energetic ultraviolet and X-ray radiation from the sun. Here, oxygen molecules absorb solar rays and warms air, making it the hottest layer. Since this layer has low air density and thus it does not feel warm, but the absorption of small amount of solar energy can cause large temperature increase. The atmosphere in the ionosphere is partially ionized and enriched ion zones exist in the form of distinct ionized layers. Since the energetic solar radiation knocks electrons off molecules and atoms, they turn them into "ions" with a positive charge.

The temperature of the thermosphere varies between night and day and across seasons, as do the numbers of ions and electrons present. So, this layer is called as the ionosphere. The ionosphere reflects the radio waves because of one or multiple reflections of short-wave radio beams from the ionized shells. As a result, long-distance radio communication is possible due to this layer.

7.4.2.5 Exosphere

"Exo" is a Greek word that means "outside external." The outer most layer of the earth's atmosphere is named as the exosphere and this layer lies between 400 and 10,000 km. At such a greater height, the density of atoms in the atmosphere is extremely low. Hydrogen and helium gases predominate in this outermost region. At an altitude of about 500–600 km, the density of the atmosphere becomes so low that collisions between the natural particles become extremely rare. The gases in this layer are very rarefied. In fact, the air in the exosphere is constant, though very gradually "leaking" out of the earth's atmosphere into outer space.

7.5 The Chemistry of Processes in the Atmosphere

7.5.1 Chemistry of the Troposphere

The chemistry of this layer is dominated by the reactions of oxygen species and oxidation reaction. Photochemical smog is produced as a result of the oxidation of organic molecules and nitrogen compounds. In spite of the high concentration of dioxygen in the atmosphere, reaction with this molecular species is not the main route toward the oxidation of trace components. The principal oxidative chemical species (in day-time chemistry) is the hydroxyl free radical, ·OH. Reactions with dioxygen require higher activation energies than reactions with free radicals such as ·OH. How fast a trace gas reacts with hydroxyl radical can be a measure of its residence time in the atmosphere. If a trace gas is inert toward the hydroxyl radicals, then its residence time will be high and it will be transported to the stratosphere, and such is the case of chlorofluorocarbons (CFCs), N_2O, CH_4, and CH_3Cl. Hydroxyl chemical reactions in the troposphere provide a very efficient scavenging mechanism for natural as well as human-made trace constituents.

At nighttime, the nitrate radical (NO_3), considered as the main oxidative chemical species under such conditions, takes over from ·OH; although NO_3 is less reactive than ·OH, it reaches its highest concentration (~400 ppt) during that period. NO_3 is easily photolyzed and therefore its daytime concentration is very low:

$$NO_2 + O_3 \rightarrow NO_3 + O_2$$

Peroxyacetyl nitrate (PAN), $CH_3COOONO_2$, is also a source of ·OH radicals when it thermally decomposes. Thus, decomposition of PAN in the presence of NO can lead to the formation of ·OH radicals and conversion of NO to NO_2 merely from thermal reactions. This could be very important because PAN can be an indicator of the urban ozone-forming potential and the organic oxidizing capacity of the urban air mass. The thermal decomposition of PAN also leads to the formation of NO_3 radicals.

7.5.2 Stratospheric Chemistry

Stratospheric ozone protects life on the surface of the earth by screening the harmful UV light coming from the sun through a photodissociation mechanism.

$$O_3 + h\upsilon \rightarrow O_2 + O$$

$$O + O_2 \rightarrow O_3$$

Thus, ozone absorbs UV radiation without being consumed; the net result is to convert UV radiation into heat. This is why temperature increases with altitude in the stratosphere, giving rise to the inversion layer that traps molecules in the troposphere. Ozone is formed if an O atom collides with an O_3 molecule:

$$O + O_3 \rightarrow 2O_2$$

"Recombination" reaction is slow, and if it were the only mechanism for ozone loss, the ozone layer would be thicker than it really is. Certain trace chemical species, mainly free radicals such as oxides of nitrogen (NO and NO_2), atomic hydrogen (H), oxygen species (OH and HO_2), and chlorine species (Cl, ClO, and ClO_2), are responsible for catalyzing the combination reaction. The thickness of the ozone layer is then the result of a competition between photodissociation and recombination.

7.5.3 Chemistry of Mesosphere

The most important minor electrically neutral constituents are compounds of oxygen, hydrogen, and nitrogen. In fact, the allotropes of oxygen are in many ways the most significant because they screen out solar ultraviolet radiation and provide the principal source of mesospheric heating (absorption of solar UV by O_2 and ozone). We also discuss oxides of nitrogen and hydrogen that strongly influence the balance of odd oxygen (O and O_3). Brief discussions of the chemistry of carbon compounds and excited species are real, so included. The chemistry of ionic species present in the mesosphere is very important because it strongly influences the propagation and absorption of radio waves. Because of ion clustering and negative-ion formation, such chemistry is extremely complex, and it is only very recently that we have begun to understand it. The current state of knowledge is discussed in some detail. The principles involved in constructing models for predicting the distribution of minor constituents, both neutral and ionic, are presented.

7.5.4 Chemistry of Ionosphere

Ionosphere can be differentiated into various layers that represent zones of different electron densities. The ionosphere, like a saw hole, is electrically neutral since it also has positive ions such as O_2^+, O^+, and NO^+. The positive ion chemistry is highly distinctive for various layers of ionosphere.

Electrons in ionosphere are produced by photoionization. Above the altitude of 100 km, this photoionization is brought about largely by extreme ultraviolet radiation. At lower altitudes, Lyman-A radiation is important. Some contribution to photoionization at somewhat lower altitudes is also made by cosmic rays. However, due to magnetic shielding of the earth, cosmic radiation is only important at fairly high latitudes. Nighttime ionization is attributed to a downward flux of protons and radiation from excited species in the upper atmosphere (i.e., UV nightglow). In the D-region of ionosphere, electrons are produced principally by photoionization of nitrous oxide, because it has lowest ionization potential among dominant species in the atmosphere. However, NO^+ is not the predominant positively charged species in the upper atmosphere. At altitude of about 80 km, the principal ion is a water cluster $H^+ (H_2O)_2$. The charge initially carried by NO^+ is transferred to water via an O_2^+ intermediate.

In the F-layer of ionosphere, positive charge is largely carried by O^+, while at lower levels, it is more likely to be present on NO^+, O_2^+, and lower down in the atmosphere on hydrated proton. The E-region of ionosphere is interesting because it sometimes shows thin sporadic layers that appear to be deriving metal-ion chemistry in mid-latitudes. Intensities of these layers show significant increases in response to meteor showers, so it is possible that metal ions have extraterrestrial origin. Typical reactions are:

$$Mg + h\upsilon \rightarrow Mg^+ + e^-$$

$$Mg^+ + O_2 + M \rightarrow MgO_2{}^+ + M$$

$$Mg^+ + O_3 \rightarrow MgO^+ + O_2$$

The first reaction produces electrons but subsequently they react with charged metal and metal oxide species.

In the ionosphere, O^+ ions are normally removed through reaction with oxygen and nitrogen:

$$O^+ + O_2 \rightarrow O_2{}^+ + O$$

$$O^+ + N_2 \rightarrow N_2{}^+ + O$$

However, reactions involving hydrogen or water are about 1000 times faster. This leads to considerable reduction in concentration of electrons through following reactions:

$$O^+ + H_2O \rightarrow H_2O^+ + O$$

$$O^+ + H_2O \rightarrow H^+ + H$$

It is followed by:

$$e^- + H_2O^+ \rightarrow H_2 + O$$

$$e^- + H_2O^+ \rightarrow OH + H$$

$$e^- + OH^+ \rightarrow O + HH$$

7.6 Climate Change: Causes and Consequences

Climate change is the periodic modification of the earth's climate brought about as a result of changes in the atmosphere as well as interactions between the atmosphere and various other geologic, chemical, biological, and geographic factors within the earth system.

Scientists attribute the global warming trend that is observed since the mid-twentieth century to the human expansion of the "greenhouse effect" that results in warming of the earth when the atmosphere traps heat radiating from the earth toward space. Greenhouse gases act like a blanket around the earth, trapping energy in the atmosphere and causing it to warm. This phenomenon is called the greenhouse effect and is natural and necessary to support life on the earth. However, the buildup of greenhouse gases can change the earth's climate and result in dangerous effects to human health and welfare and to ecosystems. The potent greenhouse gases and their relative concentrations are given in Table 7.4.

Certain gases in the atmosphere block heat from escaping. Long-lived gases that remain semipermanently in the atmosphere and do not respond physically or chemically to changes in temperature are described as "forcing" climate change. Gases, such as water vapor, that respond physically or chemically to changes in temperature are referred to as "feedbacks."

Gases that contribute to the greenhouse effect include the following:

- *Water vapor*: The most abundant greenhouse gas, but importantly, it acts as a feedback to the climate. Water vapor increases as the earth's atmosphere warms, but so does the possibility of clouds and precipitation, making these some of the most important feedback mechanisms to the greenhouse effect.
- *Carbon dioxide (CO_2)*: A minor but very important component of the atmosphere, carbon dioxide is released through natural processes such as respiration and volcanic eruptions and through human activities such as deforestation, land use changes, and burning fossil fuels. Humans have increased atmospheric CO_2 concentration by 47% since the Industrial Revolution began. This is the most important long-lived "forcing" of climate change.
- *Methane*: A hydrocarbon gas produced by both natural sources and human activities, such as decomposition of wastes in landfills, agriculture, especially rice cultivation, as well as ruminant digestion and manure management associated

Table 7.4 Pre-concentration and current concentration of GHGs

Gas	Pre-concentration—1750	Current concentration
CO_2 (ppm)	280	400.16
CH_4 (ppb)	722	1842
N_2O (ppb)	270	327
O_3 (DU)	237	337
CFC-11 (ppt)	0	236
CFC-12 (ppt)	0	527

with domestic livestock. On a molecule-for-molecule basis, methane is a far more active greenhouse gas than carbon dioxide, but also one which is much less abundant in the atmosphere.

- *Nitrous oxide*: A powerful greenhouse gas produced by soil cultivation practices, especially the use of commercial and organic fertilizers, fossil fuel combustion, nitric acid production, and biomass burning.
- *Chlorofluorocarbons (CFCs)*: Synthetic compounds entirely of industrial origin used in a number of applications, but now largely regulated in production and release to the atmosphere by international agreement for their ability to contribute to destruction of the ozone layer. They are also greenhouse gases.

The consequences of changing the natural atmospheric greenhouse are difficult to predict, but some effects seem likely:

- On average, the earth will become warmer. Some regions may welcome warmer temperatures, but others may not. Higher temperature enhances potential evaporation and duration of heat waves.
- Warmer conditions will probably lead to more evaporation and precipitation overall, but individual regions will vary, some becoming wetter and others dryer. Severe droughts have occurred over large areas, adversely affecting crop production.
- A stronger greenhouse effect will warm the ocean and partially melt glaciers and ice sheets, increasing sea level. Ocean water also will expand if it warms, contributing further to sea level rise.
- Outside of a greenhouse, higher atmospheric carbon dioxide (CO_2) levels can have both positive and negative effects on crop yields. Some laboratory experiments suggest that elevated CO_2 levels can increase plant growth. However, other factors, such as changing temperatures, ozone, and water and nutrient constraints, may more than counteract any potential increase in yield. If optimal temperature ranges for some crops are exceeded, earlier possible gains in yield may be reduced or reversed altogether.
- Climate extremes, such as droughts, floods, and extreme temperatures, can lead to crop losses and threaten the livelihoods of agricultural producers and the food security of communities worldwide. A significant decline in winter rainfall can occur, resulting in severe water scarcity during early summer months. Depending on the crop and ecosystem, weeds, pests, and fungi can also thrive under warmer temperatures, wetter climates, and increased CO_2 levels, and climate change will likely increase weeds and pests.
- Finally, although rising CO_2 can stimulate plant growth, research has shown that it can also reduce the nutritional value of most food crops by reducing the concentrations of protein and essential minerals in most plant species. Climate change can cause new patterns of pests and diseases to emerge, affecting plants, animals, and humans, as well as creating new risks for food security, food safety, and human health.

- The per capita availability of freshwater in India is expected to drop from around 1900 m^3 currently to 1000 m^3 by 2025 due to a combination of population growth and climate change. Massive floods, especially in the flood plains of the eastern Himalayan rivers, their major tributaries, and delta regions, are a major consequence of climate change. Coastal flooding and salinity intrusion from sea level rise in combination with the amplification of storm surges from more intense tropical cyclones in the Bay of Bengal can severely affect crop production.
- Swift melting of Himalayan glaciers and ice caps initially causes increased river flows, sedimentation, and subsequent reduced flow, especially in the dry months. Data from NASA's GRACE satellites and GPS stations scattered around Greenland's coast showed that between 2002 and 2016, Greenland lost approximately 280 billion tons of ice per year.
- Climate change has serious health impacts due to heat-related stress and vector-borne diseases.
- Climate change will intensify different environmental pressures and affect sustainable development.
- Due to climate change, new ocean currents are developed in the east central Pacific, affecting the climate of the area and sometimes global climate. The warm water current developed in the region is known as the EL NINO phenomenon, which affects the southern oscillation.

7.6.1 Climate Change, Climate Variability, and Global Warming

Climate change refers to a change in the state of the climate that can be identified (e.g., by using statistical tests) by changes in the mean and/or the variability of its properties, and that persists for an extended period, typically decades or longer (IPCC). In other words, climate change includes major changes in temperature, precipitation, or wind patterns, among other effects, that occur over several decades or longer.

Climate variability refers to variations in the mean state and other statistics (such as standard deviations and the occurrence of extremes) of the climate at all spatial and temporal scales beyond that of individual weather events. Variability may be due to natural internal processes within the climate system (internal variability) or variations in the nature (IPCC).

Global warming refers to the recent and ongoing rise in global average temperature near the earth's surface. It is caused mostly by increasing concentrations of greenhouse gases in the atmosphere. Global warming is causing climate patterns to change. However, global warming itself represents only one aspect of climate change. Global warming can be referred to as the long-term heating of the earth's climate system observed since the "pre-industrial" period (between 1850 and 1900) due to human activities, primarily fossil fuel burning, which increases heat-trapping greenhouse gas levels in the earth's atmosphere. The term is frequently used interchangeably with the term climate change; though the latter refers to both human and naturally produced warming and the effects it has on our planet. It is most commonly measured as the average increase in the earth's global surface temperature.

Since the "pre-industrial" period, human activities are estimated to have increased the earth's global average temperature by about 1 °C (1.8 °F), a number that is currently increasing by 0.2 °C (0.36 °F) per decade. Most of the current warming trend is extremely likely (greater than 95% probability) the result of human activity since the 1950s and is proceeding at an unprecedented rate over decades to millennia.

7.6.2 Climate Change in Jammu and Kashmir

In the context of India, especially Jammu and Kashmir, which nestles in fragile Himalayan ecosystem, the indicators of climate change are evident now. Climate change poses a serious threat to the agriculture, horticulture, water resources, tourism, species diversity, habitats, forests, wildlife, and livelihood in the region. As per the United Nations Environment Programme (UNEP) report, some parts of the state are moderate to highly vulnerable. As per the Indian Network on Climate Change Assessment (INCCA), the number of rainy days in the Himalayan region may increase by 5–10 days on an average in the 2030s, with an increase of more than 15 days in the eastern part of the Jammu and Kashmir region. The intensity of rain fall is likely to increase by 1–2 mm/day. This is likely to impact some of the horticultural crops. The rate of recession of glaciers is reportedly varying, which is being attributed to winter precipitation, climate warming, and anthropogenic elements. Temperature, precipitation, and cold wave are most likely to significantly impact the agriculture sectors.

The deficit in food production is growing in Jammu and Kashmir. With the reduction in rainfall, the rain-fed agriculture will suffer the most. Horticultural crops like apple are also showing decline in production, particularly due to decline in snowfall. About 34% and 39% of the forested grids are likely to undergo shifts in vegetation type with a trend toward increased occurrence of the wetter forest types. Climate change has impact on human health such as impacts of thermal stress, vectors, water-borne pathogens, water quality, air quality, and food availability. It is projected that the spread of malaria, tick-borne diseases etc., will all be enhanced.

The annual temperature is projected to increase from 0.9 ± 0.6 °C to 2.6 ± 0.7 °C in the 2030s. The net increase in temperature ranges from 1.7 to 2.2 °C since the 1970s. Seasonal air temperatures also show a rise in all seasons. The annual rainfall in the Himalayan region is likely to vary between 1268 ± 225.2 and 1604 ± 175.2 mm in the 2030s. The projected precipitation is likely to increase by 5–13% in the 2030s compared to the 1970s. The intensity of rain fall is likely to increase by 1–2 mm/day. The water yield in the Himalayan region, mainly covered by the Indus River, is likely to increase by 5–20% in most of the areas, with some areas in Jammu and Kashmir showing an increase of up to 50% compared to the1970s.

7.6.3 Past and Future Changes of Greenhouse Gases in the Atmosphere

The recent increase in greenhouse gases since "pre-industrial" times can be related to human activities. Climate models predict a significant global warming of several degrees within the next century if the industrial emissions increase unabated. On the other hand, accelerated policies on emission control will significantly reduce the warming after a response time of a few decades. Since the beginning of the Industrial Revolution, the levels of atmospheric greenhouse gases such as carbon dioxide (CO_2), methane (CH_4), nitrous oxide (N_2O), and tropospheric ozone (O_3) have been increasing due to human activities such as fossil fuel combustion, deforestation, and agriculture. Fossil fuel consumption during the last two centuries is the major source of the reported 25% increase in CO_2 levels compared to "pre-industrial" levels. The unabated continuation of the rate of present-day emissions would cause a CO_2 doubling, with respect to the present level, by the end of the next century. However, the climatic consequences, such as the degree of warming and the impact on life on earth, are less certain due to the complexities of the climate system and its many interactive components (atmosphere, oceans, land, biosphere, cryosphere), the majority of which are poorly understood.

7.6.3.1 Observed Climatic Variations

An apparent characteristic of the earth's climate is its variability on all time-scales from a few years to hundreds of millions of years. The most prominent example is the Pleistocene Ice-Age cycle commencing about 2,000,000 years before present time (BP) with a typical time-scale of approximately 100,000 years. During this period the global mean surface temperature varied by 5–7 °C with corresponding changes of ice volume and sea level. For most of this time the temperature was considerably lower than at present. Exceptions are a few interglacial periods, such as the so-called Eemian climate optimum around 125,000 BP.

These paleo-climatic variations have been obtained from natural records of climate-sensitive phenomena analyzed from ocean sediments, fossil corals, ice cores, tree rings, and glacier termini.

The modem instrumental record of temperature (sea surface temperature, surface air temperature) goes back to the middle of the nineteenth century. Three research groups have used these data for assessing the variation of the global mean surface temperature over the last 130–140 years. Although the instrumental data are certainly much more reliable than the proxy data, the analysis is fraught with several problems; for example, the poor spatial coverage of the measurements, changes that have occurred in observing practices, or changes in the environment that have taken place around many stations. Despite disagreements on methods for data correction, there is general agreement that the global mean temperature increased by approximately 0.5 °C from the relatively cool late nineteenth century to the relatively warm 1980s. This increase, however, has not been monotonic but caused essentially by the abrupt warming in the 1920s and during the last two decades. A more detailed review of recent climate variation may be found in IPCC.

7.6.3.2 Greenhouse Gases

The greenhouse effect life on the earth is closely connected to the concentration of atmospheric trace gases such as water vapors and carbon dioxide, often referred to as greenhouse gases. Although their concentrations are small as compared to the major atmospheric constituents, nitrogen, and oxygen, their climatic impact is substantial since they trap a significant amount of long-wave (infrared) radiative energy emitted from the surface and radiate it back to the surface.

On the basis of satellite measurements of the outgoing long-wave radiation at the top of the atmosphere, it has been estimated that nearly 150 W m^{-2} are trapped by atmospheric greenhouse gases, which is roughly 35% of the surface emission, and approximately 30 W m^{-2} are trapped by clouds. From energy balance requirements, it is easy to show that without this natural greenhouse effect the global mean surface temperature would be approximately 33 K colder than at present provided the albedo of the earth remained unchanged. The major matter of scientific and public concern is the additional greenhouse effect caused by human activities. The gradual increase in atmospheric greenhouse gases will create an imbalance between the solar radiation absorbed by the earth–atmosphere system, which remains nearly unchanged, and the reduced energy loss by increased atmospheric trapping of infrared radiation. The excess energy will heat the climate system in order to restore the energy balance. The amount of heating will depend not only on the radiative forcing by increased levels of greenhouse gases but also on the response of the climate system to the radiative forcing. The response time for heating will be governed essentially by the slow components of the climate system such as oceans and ice.

7.6.3.3 Natural Variation

One of the basic problems of climate dynamics is to understand the life cycle of atmospheric greenhouse gases and their effects on the climate. The close correspondence between trace gas abundances and climatic variation on "Ice-Age" time-scales has been documented by a wealth of ice core analyses of old polar ice extracted in Greenland and Antarctica. During glacial time, the atmospheric CO_2 content was reduced to about 70% of its "pre-industrial" value and to nearly 50% of its present value. The levels of CO_2 and CH_4 are closely correlated with the temperature variations over the last 160,000 years, as analyzed from the Vostok ice core in Antarctica. Although the resolution of the data is not sufficient to determine the phase relationship between temperatures and trace gas concentration, it is generally believed that climate change occurred first as a result of the quasiperiodic variations of the earth's orbital parameters (Milankovitch hypothesis). The shifts of CO_2 and CH_4 are probably linked with changes of the ocean circulation (CO_2) and permafrost regions (CH_4), and with the interplay of biological, chemical, and physical processes. Finally, changed levels of CO_2 and CH_4 may have enhanced the initial temperature perturbation as a result of climatic feedbacks. Although the precise nature of these interactions is not known, the data are consistent with the hypothesis of a positive feedback between climatic change and varying levels of atmospheric greenhouse gases.

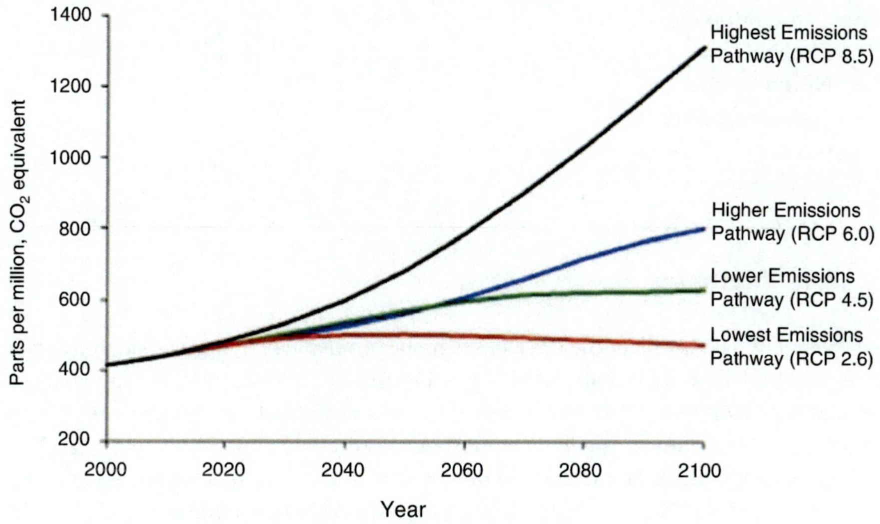

Fig. 7.2 Projected atmospheric greenhouse gas concentration

Figure 7.2 shows projected greenhouse gases concentrations for four different emissions pathways. The top pathway assumes that greenhouse gas emissions will continue to rise throughout the current century, as depicted in Fig. 7.2. The bottom pathway assumes that emissions reach a peak between 2010 and 2020, declining thereafter.

7.7 Sources and Sink of Greenhouse Gases

Greenhouse gases in the atmosphere play a critical role in shaping the global climate, and human activities have significantly modified the concentrations of many of these gases. A key area of scientific research in understanding the effects of human activities on global climate is the identification and quantification of these greenhouse gas flows.

Greenhouse gases cycle through the oceans and the biosphere over time periods that can range from a few days to millions of years. Carbon, for example, may be stored deep within ocean sediments for many millions of years or it might be cycled back into the atmosphere in a matter of hours. Scientists are trying to understand the various sources and reservoirs or sinks of each of the greenhouse gases in order to create better models of how human actions may affect natural processes. A source is any process or activity through which a greenhouse gas is released to the atmosphere. Both natural processes and human activities release greenhouse gases. A sink is a reservoir that absorbs supa chemical element or compound from another source to fit the natural cycle.

The major greenhouse gases are:

- Carbon dioxide
- Methane
- Nitrous oxide
- Halogenated species
- Ozone
- Water vapor

7.7.1 Methane

Methane (CH_4) is one of the trace gases in the atmosphere that is considered to play a major role in what is called the "greenhouse effect." Despite its still extremely minute concentration (around 1.7 ppmv), this radiatively and chemically reactive gas has been accumulating in the atmosphere at the rate of about 1% per year. Today the methane concentration is about double that in the "pre-industrial" era and two to six times higher than during the first emergence of *Homo sapiens*.

The Intergovernmental Panel on Climate Change (IPCC) estimates that 60% of total global methane emissions are related to human activities. Methane emission levels from specific sources can vary significantly, depending on factors such as climate, industrial and agricultural production characteristics, energy types and usage, and waste management practices. For example, both temperature and moisture have a significant effect on the anaerobic digestion process – a key biological process causing methane emissions in both human and natural sources.

7.7.1.1 Sources of Methane Emission

According to the available literature, there are six major sources of methane emission:

1. Anaerobic decomposition of vegetation in natural wetlands.
2. Emissions from livestock production systems that include enteric fermentation in animals and methane release from animal waste.
3. Natural (fossil) gas losses during fossil fuel exploration, processing, and distribution (coal mining, natural gas, petrochemical industry).
4. Anaerobic decomposition in paddy rice fields.
5. Biomass burning (including forest and savannah fires, burning of agricultural waste, charcoal production, and firewood combustion).
6. Anaerobic decomposition of waste in landfills.

In addition, some minor sources are discussed in the literature, such as the methane production of termites, emissions from the oceans and fresh water, and emissions from CH_4 hydrated stabilization.

The sources can be grouped into anthropogenic and natural sources. It is also possible to distinguish bacterial (so-called biogenic) from nonbacterial (pyrogenic) methane. (2), (3), (5), and (6) can be grouped into anthropogenic sources and (1) and

(4) can be grouped under natural sources, whereas (1), (2), (4), and (6) comes under bacterial source and (3) and (5) can be grouped under nonbacterial source.

Bacterial methane is considered to contribute some 80% (±10%) of the total CH_4 emission, whereas nonbacterial methane is believed to contribute 20% (±10%).

Nonbacterial methane results from thermal alteration of buried organic matter or incomplete combustion of biomass and fossil fuels. According to the 1992 "best estimate" of the CPCC committee, the contribution of the six major sources of atmospheric methane is as follows: wetlands (22%), enteric fermentation in animals and animal wastes (20%), coal mining, natural gas and petrochemical industry (19%), flooded rice fields (12%), biomass burning (8%), and landfills (6%).

The proportion of CH_4 produced from sources related to fossil carbon can be estimated from studies of the carbon-14 content of atmospheric CH_4. Three independent estimates are 21 ± 3%, 25 ± 4%, and 16 ± 12%. Emissions from wetlands are the largest natural source of CH_4 to the atmosphere.

Although the recent estimate is in the range of the IPCC (1990) estimate, the latitudinal distribution of the emissions has been revised. Emissions from high-latitude peat bogs are about half of that previously estimated, while inclusion of bubbling tends to enhance the estimates of emissions from tropical swamps. While the estimate of CH_4 emissions from landfills has been slightly reduced because of observed high CH_4 oxidation rates in landfill cover soils, two "waste" sources of CH_4 were not included in the previous IPCC assessment, i.e., emissions from animal waste and domestic sewage treatment. Based on recent studies, the annual emission from termites was scaled down to 20 Tg.

7.7.1.2 Sink

Methane in many soils can be consumed by oxidizing to carbon dioxide via methane oxidizing bacteria (methanotrophs). Although this process implies exchanges one greenhouse gas for another, methane is much more powerful than carbon dioxide as a greenhouse gas.

Hydroxyl radicals often counted as methane sinks, but technically, they do not result in methane storage or removal from the atmosphere. These radicals initiate a series of chemical reactions by which methane becomes one of several non-greenhouse compounds that are then removed from the atmosphere through precipitation or another means. Humans can also capture and utilize methane, thereby affecting overall emission levels through the use of technology in industry (such as coal mining) or waste management (landfills).

7.7.2 Halogenated Species (HFCS, PFCS, SF6)

7.7.2.1 Sources

Man is completely responsible for emissions of greenhouse gas halocarbons, many of which are synthetic chemicals being used as alternatives to ozone-depleting substances, such as CFCs. However, while halocarbons do not deplete the ozone, they are potent greenhouse gases. Sources of these gases include electrical transmission and distribution systems, semiconductor manufacturing, and aluminum and magnesium production. Like nitrous oxide, halocarbons are not stored in significant amounts through natural processes or actively taken out of the atmosphere.

Refrigeration and air conditioning sectors are recovering and recycling CFC refrigerants and increasing the use of HCFCs and ammonia. The main sources for the fully fluorinated chemicals are not well understood. Production of aluminum and use in electrical equipment are probably some of the most important sources. Antarctic macroalgae are the sources of volatile halogenated organic compounds. Twenty-eight different species of Antarctic macroalgae were collected at King George Island, South Shetlands, and it was seen that dibromomethane, bromoform, dibromochloromethane, diiodomethane, and choloroiodomethane (1991–1992) were released.

7.7.2.2 Sinks

The only significant sink for the fully halogenated CFCs is photolysis in the stratosphere, whereas the primary sink for the partially halogenated chemical is reaction with OH in the troposphere.

7.7.3 Carbon Dioxide

Carbon dioxide is odorless at normally encountered concentrations, but at high concentrations, it has a sharp and acidic odor. At such concentrations, it generates the taste of soda water in the mouth. Because carbon dioxide is soluble in water, it occurs naturally in ground water, rivers and lakes, ice caps, glaciers, and sea water. It is present in deposits of petroleum and natural gas.

7.7.3.1 Sources

There are both natural and human sources of carbon dioxide (CO_2) emissions. Natural sources include decomposition, ocean release, respiration, and volcanoes. Human sources come from activities such as cement production, deforestation, and the burning of fossil fuels. Since the beginning of the Industrial Revolution around 1750, human activities have produced a 45% increase in the atmospheric concentration of carbon dioxide from 280 ppm in 1750 to 415 ppm in 2019. Figure 7.3 shows the statistics of carbon dioxide emission both from natural and man-made sources.

7.7.3.2 Emission from Fossil Fuels

Fossil fuel (combustion of petroleum, natural gas, coal) is the primary source of CO_2 emission. The United States (North America) is currently a net source of atmospheric CO_2 from the combustion of fossil fuels. Emissions increase at the rate of 3.4% per year between 2000 and 2008. In 2017, CO_2 emissions from burning fossil fuels for energy were equal to about 76% of total US anthropogenic GHG emissions and 93% of total US anthropogenic CO_2 emission. Coal contributed 40% of the fossil fuel CO_2 emission in 2008, compared to 37% from 1990 to 2006.

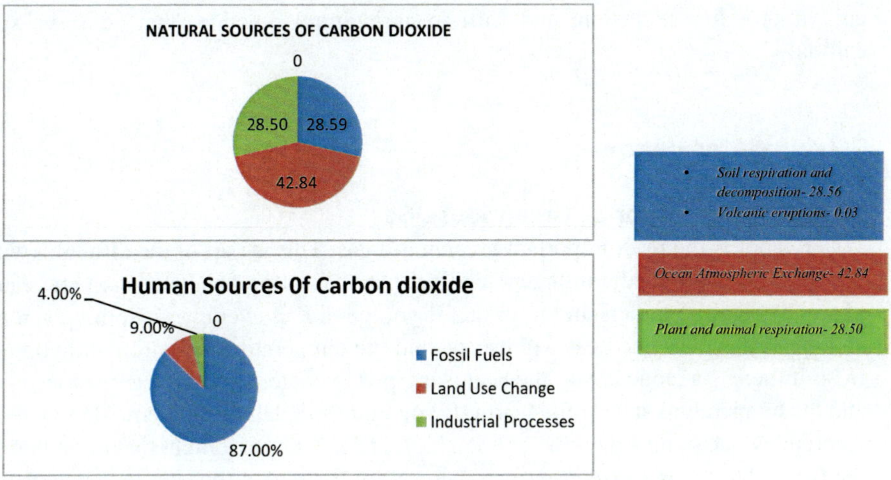

Fig. 7.3 Emissions by natural and manmade sources

7.7.3.3 Land Use Change CO_2 Emission

Emissions from land use change (LUC) are second largest anthropogenic source of CO_2. Deforestation, logging, and intensive cultivation of crop land soil emits CO_2. LUC emissions are due to both current deforestation and the carryover effects of CO_2 losses from areas deforested in previous years for the period 1990–2005. Net LUC CO_2 emissions were 1.5 ± 0.7 Pg C/year and were dominated by tropical deforestation (based on United Nations data for LUC areas). In deforestation process, fire is the primary means by which forests are converted into pastures or crop lands, after timber exploitation.

7.7.3.4 CO_2 Sink

Thirty percent of CO_2 released as global anthropogenic emission is taken up by and stored in terrestrial ecosystems around the world. The terrestrial sink is thought to be located predominantly in Northern Hemisphere known as Northern Hemisphere sinks containing both oceanic and terrestrial biosphere components and the second is weak southern sink, yet its magnitude as well as contributions are unknown.

SOCCR, one of the US Climate Change Science Program's Synthesis and Assessment Products, concluded that vegetation removed equivalent of 30% of fossil fuel-based atmospheric CO_2 emission. Half of that was contributed to regrowth of forests on agricultural land and on timber land recovering from commercial harvest.

The inventory-based and model-based estimates agreed that forests are single largest contributor to CO_2 sink. Forest regrowth in the United States is responsible for 30–70% CO_2 sink. LUC emissions are partly compensated by CO_2 uptake from regrowth of secondary vegetation and rebuilding of soil carbon pools, following afforestation, abandonment of agriculture (including the fallow phase of shifting cultivation), fire exclusion, and shift to agricultural practices that conserve soil carbon.

7.7.4 Water Vapor

7.7.4.1 Water Vapor as Greenhouse Gas

Water vapor is the most important greenhouse gas. This is part of the difficulty with the public and the media in understanding that 95% of greenhouse gases are water vapor. The public understand it, in that if you get a Fall evening or Spring evening and the sky is clear the heat will escape and the temperature will drop and you get frost. If there is a cloud cover, the heat is trapped by water vapor as a greenhouse gas and the temperature stays quite warm. If you go to InSalah in southern Algeria, they recorded at one point a daytime or noon high of 52 °C; by midnight that night, it was −3.6 °C. That was caused because there is no, or very little, water vapor in the atmosphere and it is a demonstration of water vapor as the most important greenhouse gas.

When sceptics use this argument, they are trying to imply that an increase in CO_2 is not a major problem. If CO_2 isn't as powerful as water vapor, which is abundant, adding a little more CO_2 couldn't be that bad, right? What this argument misses the fact that water vapor creates what scientists call a positive feedback loop in the atmosphere, making any temperature changes larger than they would be otherwise. How does this work? The amount of water vapor in the atmosphere exists in direct relation to the temperature. If you increase the temperature, more water evaporates and becomes vapor, and vice versa. So, when something else causes a temperature increase (such as extra CO_2 from fossil fuels), more water evaporates. Then, since water vapor is a greenhouse gas, this additional water vapor causes the temperature to go up even further, a positive feedback.

The greenhouse effect that has maintained the earth's temperature at a level warm enough for human civilization to develop over the past several millennia is controlled by non-condensable gases, mainly carbon dioxide (CO_2), with smaller contributions from methane (CH_4), nitrous oxide (N_2O), and ozone (O_3). Since the middle of the twentieth century, small amounts of man-made gases, mostly chlorine- and fluorine-containing solvents and refrigerants, have been added to the mix. Because these gases are not condensable at atmospheric temperatures and pressures, the atmosphere can pack in much more of these gases. Thus, CO_2 (as well as CH_4, N_2O, and O_3) has been building up in the atmosphere since the Industrial Revolution, when humans began burning large amounts of fossil fuel.

If there had been no increase in the amounts of non-condensable greenhouse gases, the amount of water vapor in the atmosphere would not have changed, with all other variables remaining the same. The addition of the non-condensable gases causes the temperature to increase and this leads to an increase in water vapor that further increases the temperature.

There is also a possibility that adding more water vapor to the atmosphere could produce a negative feedback effect. This could happen if more water vapor leads to more cloud formation. Clouds reflect sunlight and reduce the amount of energy that reaches the earth's surface to warm it. If the amount of solar warming decreases, then the temperature of the earth would decrease. In that case, the effect of adding more water vapor would be cooling rather than warming. However, cloud cover does mean more condensed water in the atmosphere, making for a stronger greenhouse effect than non-condensed water vapor alone; it is warmer on a cloudy winter day than on a clear one. Thus, the possible positive and negative feedbacks associated with increased water vapor and cloud formation can cancel one another out and complicate matters. The real balance between them is an active area of climate science research.

7.7.4.2 Source and Sink

The most important and most abundant atmospheric greenhouse gas is water vapor. Human sources have only a small direct influence on tropospheric water vapor concentrations, because they are largely controlled by local temperatures. The water vapor concentrations can be several percent higher over warm tropical seas and very low over freezing ice fields at the poles. As planetary warming occurs, water vapor concentration increases and adds to the warning effect.

The average water vapor concentration in the lowest stratosphere is 2.5 ppm according to the most reliable measurement. The removal of stratospheric moisture during the polar night by the formation of ice particles is the only sink mechanism that is specific for water vapor. The north to south gradient in stratospheric water vapor is due to the lowest temperatures and resulting greater effectiveness of the Antarctic sink.

7.7.5 Ozone

A triatomic very reactive form of oxygen that is an unpleasant gas with a pungent odor, a major air pollution in the lower atmosphere but a beneficial component of the upper atmosphere, and is used for oxidizing, bleaching, disinfecting, and deodorizing. Ozone is a colorless gas. Chemically, ozone is very active; it reacts readily with a great many other substances. Near the earth's surface, those reactions cause rubber to crack, hurt plant life, and damage people's lung tissues. However, ozone also absorbs harmful components of sunlight, known as "ultraviolet B" or "UV-B." High above the surface, even above the weather systems, a tenuous layer of ozone gas absorbs UV-B, protecting the living things below.

7.7.5.1 Future Levels of Stratospheric Ozone

The weight of evidence suggests that the observed middle and high-latitude O_3 losses are largely due to chlorine and bromine. Therefore, as the atmospheric abundances of chlorine and bromine increase in the future, significant additional losses of O_3 are expected. Even if the control measures of the amended Montreal Protocol were to be implemented by all nations, the current abundance of stratospheric chlorine (3.3–3.5 ppbv) is estimated to increase during the next decade, reaching a peak of about 4.1 ppbv around the turn of the century. With these increases, additional middle-latitude O_3 losses during the 1990s are expected to be comparable to those observed during the 1980s. Hence, by the year 2000, O_3 depletions are expected to be about 6% in summer and about 10% in winter. In addition, there is the possibility of incurring widespread losses in the Arctic. Enhanced levels of stratospheric sulphate aerosols from natural (e.g., Mt. Pinatubo) or anthropogenic sources could possibly lead to even greater ozone losses by increasing the catalytic efficiency of chlorine- and bromine-containing chemicals through heterogeneous chemical processes. A reduction in the peak chlorine and bromine levels, as well as a hastening of the subsequent decline of these levels, hence reducing future levels of O_3 depletion, can be accomplished in a variety of ways, including an accelerated phaseout of controlled substances and limitations on currently uncontrolled halocarbons.

7.7.5.2 Tropospheric Ozone Precursors

Carbon monoxide, non-methane hydrocarbons, and nitrogen oxides tropospheric O_3 is predicted to increase with increasing emissions of nitrogen oxide (NO), as well as increasing emissions of carbon monoxide (CO), CH_4, and non-methane hydrocarbons (NMHC), when the atmospheric abundance of NO is greater than 20–30 pptv. The magnitude of ozone changes is predicted to exhibit marked variations with latitude, altitude, and season. The differences between model calculations of O_3 increases from NO emissions are large (factor of ~3) but moderate from CH_4 emissions (−50%). The differences in the predicted spatial and temporal distributions of O_3 changes are particularly large for NO emissions but again moderate for CH_4 emissions. It should also be noted that increases in CH_4, CO, and NMHC emissions lead to reduced OH values, while increased NO emissions lead to enhanced OH levels. As a result of these opposing effects, the sign of future OH changes cannot be predicted reliably. Uncertainties in present and future tropospheric OH concentrations lead to corresponding uncertainties in the lifetimes of many tropospheric species, e.g., CH_4, HCFCs, and HFCs.

7.7.5.3 Sources and Sinks

While it is clear that CO, NMHC, and NO all have significant natural and anthropogenic sources, their budgets remain uncertain. Recent data suggests that the oxidation of HCHO in the liquid phase does not produce CO, but CO_2 directly; therefore, the estimates of CO production from hydrocarbon oxidation may need to be revised downward. It has been found that NO emissions have increased by about 30% over the period 1970–1986.

7.7.6 Nitrous Oxide

N_2O is an important long-lived greenhouse gas. In addition, it is the primary source of oxides of nitrogen in stratosphere, which play important role in controlling the abundance and distribution of stratospheric ozone.

Numerous new data confirm that that present-day atmospheric concentration of N_2O is about 310 ppbv, which is about 8% greater than the "pre-industrial" era and is increasing at a rate of 0.2–0.3% per year.

7.7.6.1 Sources and Sinks of N_2O

Biological source: N_2O is a powerful atmospheric greenhouse gas and the cause of ozone layer depletion. More than two-thirds of the emissions arise from bacterial and fungal denitrification and nitrification processes in soil.

The major contributor to biological production of N_2O in many environments is the respiratory NO reductase (NOR) found in denitrifying bacteria and in some ammonia-oxidizing organisms.

Soil as a Source or Sink for N_2O

Increased use of nitrogenous fertilizers has been thought to cause greater N_2O production by soil and natural waters. After transport to stratosphere, this additional N_2O may be converted to nitric oxide and deplete ozone layer.

The evidence for an appreciable but unidentified ground surface is sink. The soil with moisture content is less than field capacity; N_2O can be emitted from these soils. Soils are more likely to be sources than sinks of N_2O for atmosphere.

Atmospheric Source and Sink

O_2 recombination may form N_2O via its reaction with N_2. It is also conceivable that other metastable states of O_3 are accessible by optical pumping from the ground state might also be a potential source of N_2O in natural atmosphere. Additional potential atmospheric sources are NO_3, NO_2, and OH with N_2, as well as N_2 with O_2. Highly vibrational excited O_2^+ could be an important new sink of N_2O in the stratosphere. In addition, reactions of NO_2 with N_2O may be a potential sink.

Hyporheic Source and Sink of N_2O

Nitrous oxide (N_2O) is a potent greenhouse gas, with an estimated 10% of anthropogenic N_2O coming from the hyporheic zone of streams and rivers. However, there is a problem: the major atmospheric loss processes for N_2O are stratospheric photodissociation and stratospheric photooxidation. In making an accurate fine scale field measurement, it prevents understanding of processes of N_2O production and emission at bed form and flow line scales.

Bibliography

ametsoc.org (n.d.) https://glossary.ametsoc.org/wiki/Medium-range_forecast

Andreae MO (1991) Biomass burning—its history, use, and distribution and its impact on environmental quality and global climate. In: Global biomass burning-Atmospheric, climatic, and biospheric implications

Aselmann I, Crutzen PJ (1989) Global distribution of natural freshwater wetlands and rice paddies, their net primary productivity, seasonality and possible methane emissions. J Atmos Chem 8(4): 307–358

Baker-Blocker A, Donahue TM, Mancy KH (1977) Methane flux from wetlands areas. Tellus 29(3):245–250

Barnola JM, Raynaud D, Korotkevitch YS, Lorius C (1987) Vostok ice core: a 160000 year record of atmospheric CO_2. Nature 329(40):8–11

Belousov S, Berkovich L (n.d.) Environmental structure and function: climate system. Eolss.net. https://www.eolss.net/sample-chapters/C01/E4-03-04-06.pdf

Bingemer HG, Crutzen PJ (1987) The production of methane from solid wastes. J Geophys Res Atmos 92(D2):2181–2187

Bonan GB (2008) Forests and climate change: forcing, feedbacks, and the climate benefits of forests. Science 320:1444–1449

Boutron CF, Görlach U, Candelone JP, Bolshov MA, Delmas RJ (1991) Decrease in anthropogenic lead, cadmium and zinc in Greenland snows since the late 1960s. Nature 353(6340):153–156

Broccoli AJ, Manabe S (1987) The influence of continental ice, atmospheric CO_2 and land albedo on the climate of the last glacial maximum. Climate Dynamics 1:87–99

Butler JM, Nolan DJ, Vertes EL, Varnum-Finney B, Kobayashi H, Hooper AT, Seandel M, Shido K, White IA, Kobayashi M, Witte L (2010a) Endothelial cells are essential for the self-renewal and repopulation of Notch-dependent hematopoietic stem cells. Cell Stem Cell 6(3): 251–264

Butler MP, Davis KJ, Denning AS et al (2010b) Using continental observations in global atmospheric inversions of CO_2: North American carbon sources and sinks. Tellus B 62:550–572

Cahir JJ (2019) Weather forecasting. In: Encyclopedia Britannica https://www.britannica.com/science/weather-forecasting

Canadel JG, 2007. Terrestrial Eco systems in a Changing World, Springer Cham

Canadell JG, Le Quere C, Raupach MR et al (2007) Contributions to accelerating atmospheric CO_2 growth from economic activity, carbon intensity, and efficiency of natural sinks. Natl Acad Sci USA 104:18866–18870

Casada ME and Safley LM (1990) Global methane emissions from livestock and poultry manure

CCSP (Climate Change Science Program) (2007) The first state of the carbon cycle report (SOCCR): the North American carbon budget and implications for the global carbon cycle. In: King AW, Dilling L, Zimmerman GP et al (eds) A report by the US Climate Change Science Program and the Subcommittee on Global Change Research. NOAA, National Climatic Data Centre, Asheville, NC

ChallinorA J, Watson J, Lobell DB, Howden S, Smith D, Chhetri N (2014) A meta-analysis of crop yield under climate change and adaptation. Nature Climate Change 4(4):287–291

Chapman L, Thornes JE (2003) The use of geographical information systems in climatology and meteorology. Prog Phys Geogr 27:313–330

Chappellaz J, Barnola JM, Raynaud D, Korotkevich YS, Lorius C (1990) Ice-core record of atmospheric methane over the past 160,000 years. Nature 345(6271):127–131

Chappellaz J, Barnola JM, Raynaud D, Korotkevich YS, Lorius C (1993) Ice-core record of greenhouse gases. Science 259(5097):926–934

Clark PU, Shakun JD, Baker PA, Bartlein PJ, Brewer S, Brook E, Carlson AE, Cheng H, Kaufman DS, Liu Z, Marchitto TM (2012) Global climate evolution during the last deglaciation. Proc Natl Acad Sci 109(19):E1134–E1142

Clarke L, Edmonds J, Jacoby H, Pitcher H, Reilly J, Richels R (2007) CCSP synthesis and assessment product 2.1, Part A: scenarios of greenhouse gas emissions and atmospheric concentrations. US Government Printing Office, Washington DC

Conrad R, Rothfuss F (1991) Methane oxidation in the soil surface layer of a flooded rice field and the effect of ammonium. Biol Fertil Soils 12(1):28–32

Crutzen PJ, Aselmann I, Seiler W (1986) Methane production by domestic animals, wild ruminants, other herbivorous fauna, and humans. Tellus B 38(3–4):271–284

Curtis PG, Slay CM, Harris NL, Tyukavina A, Hansen MC (2018) Classifying drivers of global forest loss. Science 361:1108–1111

Denman KL, Brasseur G, Chidthaisong A, Ciais P, Cox PM, Dickinson RE, Hauglustaine D, Heinze C, Holland E, Jacob D, Lohmann U, Ramachandran S, da Silva Dias PL, Wofsy SC, Zhang X (2007a) Couplings between changes in the climate system and biogeochemistry. In: Climate change 2007: the physical science basis. Contribution of Working Group I to the Fourth Assessment Report of the Intergovernmental Panel on Climate Change. Cambridge University Press, Cambridge

Denman KL, Basseur G, Chidthaisong A (2007b) Couplings between changes in the climate system and bio-geochemistry. In: Solomon S, Qin D, Manning M et al (eds) Climate change 2007: the physical science basis. Contribution of Working Group I to the Fourth Assessment Report of the Intergovernmental Panel on Climate Change. Cambridge University Press, New York, NY

Department of Animal and Plant Sciences, University of Sheffield, Sheffield, UK

Dyras I, Dobesch H, Grueter E, Perdigao A, Tveito O, Thornes JE, van der Wel F, Bottai L (2005) The use of geographic information systems in climatology and meteorology: COST 719. Meteorol Appl 12:1–5

Environment of Earth (n.d.). http:///environmentofearth.wordpress.com/2009/09/13/chemistryofionosphere/

Farooq M (2016) Climate change & concerns of J & K. J&K Climate Change Centre

Hameed S, Dignon J (1992) Global emissions of nitrogen and sulfur oxides in fossil fuel combustion 1970–1986. J Air Waste Manage Assoc 42(2):159–163

Hansen J, Lebedeff S (1987) Global trends of measured surface air temperature. J Geophysical Res 92(133):45–72

Heilig GK (1994a) The greenhouse gas methane (CH_4): sources and sinks, the impact of population growth, possible interventions. Population Environ 16(2):109

Heilig GK (1994b) The greenhouse gas methane (CH_4): sources and sinks, the impact of population growth, possible interventions. Population and Environment 16(2):109

Hobbs PV (2000) Introduction to atmospheric chemistry. Cambridge University Press, Cambridge

Holloway AM, Wayne RP (2010) Atmospheric chemistry. RSC Publishing, Cambridge

Houghton JT, Jenkins GJ, Ephraums JJ (1990) Climate change: the IPCC scientific assessment. Am Sci 80(6)

https://www.usgs.gov/faqs/what-remote-sensing-and-what-it-used?qt-news_science_products=0

Ibanez, Jorge, Hernandez-Esparza, Margarita, Doria-Serrano, Carmen, Fregoso-Infante, Arturo, Singh, Mono (2007). The chemistry of processes in the atmosphere. http:///www.researchgate.net/publication/315525251

Imbrie J, Imbrie KP (1979) Ice Ages—solving the mystery. Macmillan, London

Imhoff M, Zhang P, Wolfe R, Bounoua L (2010) Remote sensing of the urban heat Island effect across biomes in the continental USA. Remote Sens Environ 114:504–513

Indian Meteorological Department Pune (n.d.) https://www.imdpune.gov.in/Clim_Pred_LRF_New/FAQ.html

Indian Network for Climate Change Assessment (n.d.) Climate change and India: a 4×4 assessment. A sectoral and regional analysis for 2030s. http://gbpihed.gov.in/PDF/Publication/Climate_change_4_4.pdf

IPCC (1990) Climate change-the IPCC scientific assessment. In: Houghton JT, Jenkins GJ, Ephraums JJ (eds) Report prepared for IPCC by Working Group I. Cambridge University Press, Cambridge

IPCC, 2012: Glossary of terms. In: Managing the risks of extreme events and disasters to advance climate change adaptation Field, C.B, V Barros, TF Stocker, D. Qin, D.J. Dokken, K.L. Ebi, M.D. Mastrandrea, K.J. Mach, G.K. Plattner, S.K. Allen, M. Tignor, and P.M. Midgley. A special report of Working Groups I and II of the Intergovernmental Panel on Climate Change (IPCC). Cambridge University Press, Cambridge, pp. 555–564

IPCC Fifth Assessment Report (2009) United States global change research program. In: global climate change impacts in the United States. Cambridge University Press, Cambridge

Jackson ST (2021) Climate change. In: Encyclopedia Britannica https://www.britannica.com/Science/climate-change

Jedlovec GJ, Lerner JA, Atkinson RJ (2000) A satellite-derived upper-tropospheric water vapor transport index for climate studies. J Appl Meteorol 39(1):15–41

Jin M (2004) Analysis of land skin temperature using AVHRR observations. Bull Am Meteorol Soc 85:587–600

Jin M, Shepherd JM (2005) Inclusion of urban landscape in a climate model: how can satellite data help? Bull Am Meteorol Soc 86:681–689

jkenvis.org (n.d.) http://jkenvis.org/pdf/newsletters/October_December_2015.pdf

Jones PD, Raper SCB, Bradley RS, Diaz HF, Kelly PM, Wigley, T.M.L. (1986) Northern Hemisphere surface air temperature variations. J Clim Appl Meteorol 25(16):1–79

Kashmirlife.net (n.d.) https://kashmirlife.net/climate-change-2030-1058/

Keller M, Jacob DJ, Wofsy SC, Harriss RC (1991) Effects of tropical deforestation on global and regional atmospheric chemistry. Clim Chang 19(1):139–158

Khalil MAK, Rasmussen RA, French JRJ, Holt JA (1990) The influence of termites on atmospheric trace gases: CH_4, CO_2, $CHCl_3$, N_2O, CO, H_2, and light hydrocarbons. J Geophys Res Atmos 95 (D4):3619–3634

Khalil N, O'Connor RN, Unruh HW, Warren PW, Flanders KC, Kemp A, Bereznay OH, Greenberg AH (1991) Increased production and immunohistochemical localization of transforming growth factor-b in idiopathic pulmonary fibrosis. Am J Respir Cell Mol Biol 5:155–162

King TS, Chinchilli VM, Wang KL, Carrasco JL (2007a) A class of repeated measures concordance correlation coefficients. J Biopharm Stat 17(4):653–672

King AW, Dilling L, Zimmerman GP et al (2007b) Executive summary. In: King AW, Dilling L, Zimmerman GP et al (eds) The first state of the carbon cycle report (SOCCR): the North American carbon budget and implications for the global carbon cycle. NOAA, National Climatic Data Centre, Asheville, NC

Lelieveld J, Crutzen PJ (1991) The role of clouds in tropospheric photochemistry. J Atmos Chem 12(3):229–267

Lillesand T, Kiefer R, Chipman J (2004) Remote sensing and image interpretation. John Wiley, Chichester

Manning P (1990) Slavery and African life: occidental, oriental, and African slave trades, vol 67. Cambridge University Press, Cambridge

Mathews E, Fung I (1987) Methane emission from natural wetlands: global distribution, area, and environmental characteristics of sources. Global Biogeochem Cycles 1:61–86

McNicol IM, Mand RC, Mitchard ET (2018) Carbon losses from deforestation and widespread degradation offset by extensive growth in African wood lands. Nat Commun 9:3045

Mendelsohn R, Kurukulasuriya P, Basist A, Kogan F, Williams C (2007) Climate analysis with satellite versus weather station data. Clim Chang 81:71–83

Menon S, Denman KL, Brasseur G, Chidthaisong A, Ciais P, Cox PM, Dickinson RE, Hauglustaine D, Heinze C, Holland E, Jacob D (2007) Couplings between changes in the climate system and biogeochemistry (No. LBNL-464E). Lawrence Berkeley National Lab (LBNL), Berkeley, CA (United States)

Murphy BF, Timbal B (2008) A review of recent climate variability and climate change in southeastern Australia. Int J Climatol 28(7):859–879

Myhre G, Stordal F, Restad K, Isaksen IS (1998) Estimation of the direct radiative forcing due to sulfate and soot aerosols. Tellus B 50(5):463–477

Nakicenovic N, Swart R (2000) Special report on emissions scenarios. Cambridge University Press, Cambridge

NASA Global Climate Change (n.d.-a) The causes of climate change. https://climate.nasa.gov/causes/

NASA Global Climate Change (n.d.-b) Global warming vs climate change. https://climate.nasa.gov/resources/global-warming-vs-climate-change/

Nicholls RJ (1995) Coastal megacities and climate change. Geo J 37(3):369–379

Oreskes N (2004) The scientific consensus on climate change. Science 306(5702):1686. https://doi.org/10.1126/science.1103618

Poppoff IG, Whitten RC (1976) The mesosphere. In: Geological surveys, vol 2. National Aeronautics and Space Administration, California

Purves DW, Lichstein JW, Pacala SW (2007) Crown plasticity and competition for canopy space: a new spatially implicit model parameterized for 250 North American tree species. PLoS One 2(9):e870

Quay R (2010) Anticipatory governance: a tool for climate change adaptation. J Am Plan Assoc 76(4):496–511

Raupach M et al (2007) Global and regional drivers of accelerating CO_2 emissions. Proc Natl Acad Sci U S A 104:9913–9914

Raval A, Ramanathan V (1989) Observational determination of the greenhouse effect. Nature 342: 758–761

Roeckner E (1992) Past, present and future levels of greenhouse gases in the atmosphere and model projections of related climatic changes. J Experimental Bot 43(253):1097–1109

Science Daily (n.d.) Weather forecasting. https://www.sciencedaily.com/terms/weather_forecasting.htm

Seller MJ, Nevin NC (1984) Periconceptional vitamin supplementation and the prevention of neural tube defects in south-east England and Northern Ireland. J Med Genet 21(5):325–330

P. D. Sharma, 2018 Ecology and environment, Rastogi Publications Meerut; 2 34–35

Stauffer B, Oeschger H (1985) Gaseous components in the atmosphere and the historic record revealed by ice cores. Annu Glaciol 7:54–59

Helbrough E, McGuire R (2006) The big book of knowledge. Miles Kelly Publishing, pp 128–129

Tipping E, Woof C, Rigg E, Harrison AF, Ineson P, Taylor K, Benham D, Poskitt J, Rowland AP, Bol R, Harkness DD (1999) Climatic influences on the leaching of dissolved organic matter from upland UK moorland soils, investigated by a field manipulation experiment. Environ Int 25(1):83–95

Tomlinson CJ, Chapman L, Thornes JE, Baker C (2011) Remote sensing land surface temperature for meteorology and climatology: a review. Meteorol Appl 18(3):296–306. https://doi.org/10.1002/met.287

Trenberth KE, Fasullo JT, Kiehl JT (2009) Earth's global energy budget. Bull Amer Meteor Soc 90: 311–323

U.S. DOE (2008) Carbon cycling and bio sequestration: integrating biology and climate through systems science; Report from the March 2008 Workshop, DOE/SC-108. U.S. Department of Energy Office of Science, Washington

U.S. Environmental Protection Agency (2019) Inventory of U.S Greenhouse Gas Emissions and Sinks 1990–2017, Executive Summary, April 2019, includes U.S. Territories

U.S. Environmental Protection Agency (n.d.) Climate Impacts on Agriculture and Food Supply

U.S. Geological Survey (n.d.) Volcanic Gases and Climate Change Overview. http://volcanoes.usgs.gov/hazards/gas/climate.php. Accessed 2 Aug 2014

United Nation Framework Convention on Climate Change (2011) Fact sheet: climate change science—the status of climate change science today. https://unfccc.int/files/press/backgrounders/application/pdf/press_factsh_science.pdf

Vinnikov KY, Groisman P, Lugina KM (1990) The empirical data on modem global climate changes (temperature and precipitation). J Climate 3:662

wamis.org (n.d.) http://www.wamis.org/agm/meetings/walcs10/S3-Mukhopadhyay.pdf

RT Watson, LG Meira Filho, E Sanhueza, A Janetos (1990) Chapter 1 of 1990 IPCC Scientific Assessment

Watson RT, Meira Filho LG, Sanhueza E, Janetos A (1992) Greenhouse gases: sources and sinks. Climate Change 92:25–46

Whalen CK, Henker B, Buhrmester D, Hinshaw SP, Huber A, Laski K (1989) Does stimulant medication improve the peer status of hyperactive children? J Consulting Clin Psychol 57(4): 545

Winckler J, Hand RC, Pongratz J (2017) Robust identification of local bio geophysical effects of land-cover change in a global climate model. J Clim 30:1159–1176

Hoegh-Guldberg O, Jacob D, Bindi M, Brown S, Camilloni I, Diedhiou A, Djalante R, Ebi K, Engelbrecht F, Guiot J and Hijioka Y (2018) Impacts of 1.5 °C global warming on natural and human systems. Global warming of 1.5 °C. An IPCC special report

Climate Change Impact on Plants

8

Abstract

Climate change instigated by increasing greenhouse gases from human emissions is likely to affect crops diversely from region to region. Climate change influences both agricultural crop production as well as productivity. Climate change impacts the physiology of the plants, thus affecting the overall yield and production. It can also alter plant growth and plant structure. Climate change also affects crop pests and disease susceptibility, which in turn affects crop health. In this chapter we will explore the impact of climate change on the physiology and disease susceptibility of the agricultural production or vegetation.

Keywords

Climate change · Resistance · Susceptibility · Scorching · Frost injury · Viability

Agriculture is the backbone of Indian economy, which in turn relies on the monsoon season. Rising global temperature is not only causing climate change but also contributing to the irregular rainfall patterns. Uneven rainfall patterns, increased temperature, and elevated CO_2 content in the atmosphere are important climatic parameters that affect the crop production. Research studies indicate that weathering parameters influence strongly (67%) compared to other factors such as soil and nutrient management (33%) during the cropping season. The Intergovernmental Panel on Climate Change (IPCC) projected that the global mean surface temperature will likely rise and may result in uneven climatic changes. This rising temperature may affect crop yield at large scale. It has been reported over the twentieth century that rising temperature plays an important role on global warming as compared to precipitation. Researchers have confirmed that crop yield falls by 3–5% for every 1 ° F increase in the temperature. In India, crop production may be divided into two seasons: Kharif (influenced by south-west monsoon) and Rabi (mostly influenced by

L. Ahmad et al., *Climate Change and Agrometeorology*,
https://doi.org/10.1007/978-981-99-4863-5_8

north-east monsoon). The present study shows that the crop production is dependent on temperature. Temperature vs. crop production shows a funnel shape for all the seasons. For the lower temperature, both the properties are almost linearly correlated. In Rabi, production at the beginning show a negative trend with temperature, which slowly converts to the positive trend. In Kharif, the negative trend is not visible. At higher temperatures, production increases for both the seasons but with large scattering. The findings may be helpful to study the effect of climate change on the crop production.

Climate change might decrease:

- the chilling and inhibit bloom and fruit set in horticultural crops,
- lead to high temperature and wind during bloom or ripening that could negatively impact fruit set or fruit quality,
- increase ET rates that could lead to water deficits, and
- increase problems with heat stress.

Some possible climate change impacts on agriculture include:

- droughts, floods, and other weather disasters,
- pollination affected by rainfall and other extreme events,
- faster phenological development,
- inadequate chilling requirements,
- frost and chill damage,
- the spread of new insects and diseases, and
- lower or higher yield and quality due to warming and water relations during summer. One example of warming impacts is decreased winter chilling, which leads to bad pollination, staggered bloom, reduced fruit set, and poor fruit quality. In low chill years, apricots and cherries can drop flower buds or not produce a crop when chilling is inadequate.

8.1 Global Scenario

Climate change could be one of the affecting parameters all over the world. It is predicted by IPCC that many of the observed changes due to climate change are unprecedented. Global sea level rise is projected to be between 0.17 and 0.41 m in the year 2050. It is observed that the rate of rising sea level has been larger than the mean rate during the previous two millennia till the mid-nineteenth century. IPCC reported that changes in precipitation will be nonuniform and its extreme events over most of the mid-latitude and wet tropical regions will become more intense and frequent. Recent finding of increasing trends in extreme precipitation leads to imply greater risks of flooding at regional scale. Since 1850, last three decades has been consecutively warmer than any other decade on the earth's surface. Heat wave frequency has increased since the middle of the twentieth century in large part of Asia. Moreover, concentration of CO_2 and other greenhouse gases leads to increase the temperature. IPCC report states that the amount and rate of warming expected for the twenty-first century depends on the total amount of greenhouse gases that mankind emit. These observed changes are responsible for varying the climate at different parts of the earth and sometimes it may result in extreme weather events.

8.2 Indian Scenario

Like other countries, India has also started experiencing extreme weather events that lead to change in climate. As mentioned earlier, global warming is one of the major affecting parameters to change the climate. In India, it is observed that the annual mean temperature has increased at the rate of 0.42 °C. Indian agriculture system is based on south-west and north-east monsoon. Almost 80% of the total precipitation comes from south-west monsoon in India. Any fluctuations and uncertainties in long range rainfall pattern may affect the agriculture sector and also lead to increase the frequency of droughts and floods at regional scale. A significant increasing trend in rainfall was reported along the west coast, north Andhra Pradesh, and Northwest India, while significant decreasing trend was observed over parts of Gujarat, Madhya Pradesh, and adjoining area, Kerala and Northeast India. North-western region of India gets affected by western disturbances at small scale as such disturbances have impact only on Rabi production for not more than 20–30 days. Not only monsoon but temperature has also shown its effect on agriculture. Extreme maximum and minimum temperature showed an increasing trend in the southern part, whereas decreasing trend in the northern part of India. Research studies show that with the increase in temperature, crop productivity is likely to decrease in future. Hence, there is a need to study the dependency of temperature on crop productivity, stability, yield, and quality to lift the country's economy.

8.3 Effect of Increasing Temperature on Indian Agriculture

Research studies shows that rise in global surface temperature would affect Indian agriculture. Several climatic factors that affect agriculture productivity are heat waves, high temperature, heavy and prolonged precipitation, and excess cold. These factors have positive as well as negative effects on crop production. Almost every year India faces several weather events due to changes in such climatic parameters in various regions, which reduce crop yield. Varied nature of such weather events tends to affect the crop growth cycle and plant physiological processes. In India, about 17% of the years during 1901–2010 were reported as drought years, resulting in severe impacts on agriculture, water resources, food security, economy, and social life in the country. The variation in temperature and precipitation above threshold value may affect photosynthesis and transpiration process in crops. Excess rainfall and flood may lead to physical damage of the crops. Studies predicted that changing trends in temperature and precipitation will continue to have significant impact on agriculture. A small rise in temperature (1–2 °C), especially in the seasonally dry tropical regions, would decrease crop yield.

Indian agriculture is divided into two main seasons, Kharif and Rabi, based on the monsoon. It is reported that overall temperature rise is likely to be much higher during winter (Rabi) rather than in rainy season (Kharif). Moreover, it is predicted that the mean temperature in India will rise by 0.4–2.0 °C in Kharif and 1.1–4.5 °C in Rabi by 2070. The decline in agricultural productivity leads to increase food prices at state as well as country levels. Hence, temperature could be one of the significant affecting factors, resulting in greater instability in agriculture of India.

8.4 Effect of Increased CO_2 Levels on Agriculture

8.4.1 Direct Effect

An increase in the CO_2 level may result in an increase in food production. It is predicted that a twofold increase in CO_2 will lead to a 10–15% increase in dry matter production if all other factors remain constant. As C_3 plants respond much more to an increase in the CO_2 level than do C_4 plants, crops in Central and Northern Europe and similar latitudes are expected to perform better than crops in tropical areas, where maize, sorghum, sugarcane, and millets are staples and grown in abundance. Generally, it is assumed that increased atmospheric CO_2 could enhance growth rates of certain types of crop plants and that change in temperature and precipitation would affect livestock crops, pest, and soils. Plant growth is generally limited by carbon sink rather than carbon source. Temperature-sensitive plant developmental processes may constrain many agricultural yields. In order to draw an idea about crop response under enhanced CO_2 concentration, other factors such as temperature and precipitation must be taken into consideration as their combined effect may be somewhat different. Increases in CO_2 and other greenhouse gases (GHGs) result in climate changes such as increased temperature, more erratic pattern of rainfall, and weather change. The uneven monsoon rainfall in India leads to large-scale droughts and floods, having a major effect on Indian crop production and on the economy of the country. This certainly has an impact on agriculture and this impact manifests itself in a number of ways, namely, changes in the length of growing season. Scientists suggest that a 1 °C increase in average temperature would tend to advance the thermal limit of cereal cropping in the mid-latitude Northern Hemisphere and would bring more land under cultivation than what it is today under cool climates.

8.4.2 Pests and Diseases Susceptibility

There is an intimate relationship between occurrence and distribution of pests and diseases and the temperature, and any change in temperature will have a significant effect on pest and disease development and their interaction with agricultural crops. A number of effects of global warming on insects, pests, and disease-causing organisms have been identified. These are increases in the number and rate of development in a season, expansion of area of distribution, earlier establishment in their population in a favorable season, and more intense attacks, particularly by midnight and exotic species. Again, with the shifting agricultural production to new areas and changing of the agroclimatic regions, the emergence of new insect species is very possible, posing an additional threat to agricultural production.

8.4.3 Climate Change and Availability of Water

Changes in rainfall due to global climate change may affect the surface moisture availability, which becomes important for crop stand establishment in the rain-fed areas. Modelling techniques are not reliable enough to predict precipitation changes and one can expect some increased drought in some regions and increased rainfall in others. Small changes in precipitation can have magnified run-off effects. Doubling of CO_2 predicts an 8–15% increase in global precipitation, with a 30% increase in the water-holding capacity, but the atmosphere precipitation will not keep up with potential evaporation. The problem will be more acute at higher altitudes. Run-off is also likely to decrease due to more arid conditions and the increased frequency of droughts.

8.4.4 Climate Change and Soil Fertility

Soil fertility is very essential for sustaining productivity in the world and depends on a complex network of soil structure, water, oxygen, and nutrient availability. The soil's organic matter enables it to support plant life and soil microbes. Soil microbes facilitate the decomposition of organic matter from litter fall and CO_2 is a natural by-product of this process. Rising atmospheric temperatures and/or CO_2 levels are likely to increase photosynthesis and plant productivity. However, the effects of warming on soil's molecular composition have been poorly studied. It is therefore unclear to what extent the carbon-containing components of soil matter will accumulate or degrade and thus how much carbon will be sequestered by the soil and how much will be released into the atmosphere as CO_2. The doubling of CO_2 increases plant biomass production, soil water-use efficiency by the plants, and C/N ratios of plants. The change in the C/N ratios of plant residues have impact on soil microbial processes and affect the production of trace gases NO_2 and N_2O. Higher soil temperature stimulates microbial respiration and decomposition of organic matter. Plants may take in more nutrients. Brackish-water inundation on coastal areas gives rise to potential acid sulphate soil layers.

8.4.5 Climate Change and Livestock

Climate change can impact the amount and quality of produce, reliability of production, and the natural resource base on which livestock production depends. Climate is an important factor of agricultural productivity, and climate change is expected to severely impact livestock production systems. Furthermore, global demand for animal protein will rise as populations become more affluent and eating habits change. Therefore, animal production plays (and will continue to do so) a key role in the food supply chain. While the increasing demand for livestock products offers market opportunities and income for small, marginal, and landless farmers, livestock production globally faces increasing pressure because of negative environmental implications, particularly because of greenhouse gas (GHG) emissions. Agriculture is an important sector to address as it both impacts climate as well as is influenced by climate change. Higher temperatures, potentially caused by GHG, would likely result in a decline in dairy production, reduced animal weight gain, reproduction, and lower feed-conversion efficiency in warm regions. Incidence of diseases among livestock and other animals are likely to be affected by climate change, since most diseases are transmitted by vectors such as ticks and flies (development stages of ticks and flies are often dependent on ambient temperature). Cattle, goat, horses, and sheep are also vulnerable to an extensive range of nematode worm infections, most of which have their development stages influenced by climatic conditions. Climate change will have far-reaching consequences for dairy, meat, and wool production systems that rely primarily on grass and rangelands, and this will likely detrimentally affect vulnerable pastoral communities engaged in extensive livestock production systems in dry lands. Although the direct effects of climate change on animals are likely to be small (as long as temperature increases do not exceed 3 °C), climate change will affect animals indirectly through physiological stress and thermoregulatory control, nutrition, and disease stress. Because livestock products are an incredibly important human food, and because animal farming is a significant source of income for millions of farmers, it is necessary to identify climate change mitigation strategies and solutions.

8.5 Mitigation of Climate Change

The atmospheric levels of GHGs such as CO_2, CH_4, N_2O, O_3, and chlorofluorocarbons (CFCs) are increasing with rapid industrialization and intensive agricultural and related activities. The levels of CO_2, the most important GHG, are expected to increase by 305 in the next 50 years. The more buildup of these gas levels would increase the temperature of the earth and affect climatic changes. The overall levels of production can be maintained through a combination of shifts in agricultural zones and adjustments in technology and management. Every effort should be made to arrest deforestation and promote upgrading of degraded land through agroforestry and other appropriate forms of land use. This will help to increase carbon fixation on the earth. Establishment of crop–weather watch groups, special groups consisting of meteorologists, agricultural research and extension workers, developmental administrators, and mass media representatives, should be established in every agroecological region of the country to continuously monitor the weather situation and analyze its implications for crop growth and pest incidence. Such crop–weather watch groups can benefit at the national level from the World Climate Programme of the World Meteorological Organization (WMO) and early warning system of the Food and Agriculture Organization (FAO). This will help to equip farmers and fishermen with location-specific information to increase their preparedness for floods, tropical cyclones, and drought. The loss in farm-level net revenue is estimated to range between 9% and 25% for a temperature rise of 2–3.5 °C. With the knowledge gathered to date, it is possible to consolidate the gains and adapt to changes in order to tackle the negative effects of climate change on rice production and productivity for assuring the food security of the Indian population. Some of the possible options include the following.

- **Varietal adaptation**: Considerable variation exists between the rice varieties in tolerance to high temperatures. If the sensitivity of spikelet sterility to temperature is increased by 2 °C for the new varieties, it can offset the detrimental effect completely. Two possible adaptations are likely to occur: one could be the use of varieties more tolerant to temperature in the low-latitude region and the other is the use of late maturing varieties to take advantage of the longer growing season in high-latitude areas.
- **Adjustment of planting date**: Adjusting planting dates could be another strategy, which is likely to be adopted in the future by rice farmers. At high latitudes, a rise in temperature would lengthen the period in which rice can be grown. In northern China (Shenyang), yield increases of 43% are expected by advancing the planting date of rice by 30 days. At Madurai, India, significant yield decrease was predicted if current planting dates were used under the Geographic Information System (GIS) scenario due to high spikelet sterility. It could be possible to offset these yield reductions if planting was delayed by 1 month.
- **Environment-friendly cultivation practice**: It would be useful to standardize methods of reducing the contribution of agriculture to GHG accumulation in the atmosphere either through efficient water management to control CH_4 release or proper utilization of fertilizer-N that could contribute to the emission of N_2O.

8.5.1 Plant Sense and Response to Change in CO_2 Concentrations

Despite recent interest in the effects of high CO_2 on plant growth and physiology, very little is known about the mechanism by which plants sense changes due to the concentration of this gas. Since atmospheric CO_2 concentration is relatively constant and the conductance of the cuticle to CO_2 is low, sensory mechanism is likely to exist only for intercellular CO_2 concentration. Therefore, responses of plant to changes in atmospheric CO_2 will depend on the effect of these changes on intercellular CO_2 concentration. Although a variety of plant responses to atmospheric CO_2

concentration have been reported, most of these can be attributed to the effects of intercellular CO_2 on photosynthesis or stomatal. Short-term and long-term effects of CO_2 are discussed as sensory mechanism for response of plants to atmospheric CO_2. The available data suggest that plants do not fully realize the potential increases in productivity associated with increased atmospheric CO_2. This may be because of genetic and environmental limitations to productivity or because plant responses to CO_2 have evolved to cope with variations in intercellular CO_2 caused by factors other than changes in atmospheric CO_2. Calculation showed that reduction in stomatal conductance by doubled CO_2 could potentially reduce the effects of ambient O_3 and SO_2 by 15%. Understanding how plants respond and might be adapted to a future increase in CO_2 will also help us understand how they are currently responding and how they may have adapted to the increase that has already occurred.

A workshop supported by the Office of Carbon Dioxide Effects Research, OAES, U.S. Department of Energy identified 31 plant responses to CO_2 concentration are:

- Photosynthesis increase
- Growth rate change (dry weight and elongation)
- Transpiration decrease
- Stomatal conductance decrease
- Tolerance to atmospheric pollutants increase
- Leaf increase
- Leaf dry weight/leaf area in increase
- Lateral branches increase in number
- Root/shoot ratio changes
- Fruit size, fruit number, seed per plant increases
- Inter- and intraspecific differences in responses

One of the most consistent effects of elevated atmospheric CO_2 on plants is an increase in the rate of photosynthetic carbon fixation by leaves. CO_2 is also important in regulating the openness of stomatal pores through which plants exchange gases, with the external environment. Open stomata allow CO_2 to diffuse into leaves for photosynthesis, but also provide a pathway for water to diffuse out of leaves. Plants therefore regulate the degree of stomatal opening (known as stomatal conductance) as a compromise between the goals of maintaining high rates of photosynthesis and low rates of water loss. Across a variety of Free-Air Carbon dioxide Enrichment (FACE) experiments, growth under elevated CO_2 decreases stomatal conductance of water by an average of 22%. The elevated CO_2 leads to large variety of secondary effects on plant physiology with increase in dry matter production on average by 17% for the above ground and more than 30% for below ground portions of plant. This increased growth is also reflected in the harvestable yield of crops, with wheat, rice, and soybean all showing increase in yield on average by 12–14%.

Elevated CO_2 also leads to changes in the chemical composition of plant tissue. Due to increased photosynthetic activity, the nonstructural carbohydrates (sugars and starches) per unit leaf area increase on average by 30–40% under FACE-elevated CO_2, whereas tissue nitrogen content decreases. Protein concentrations in grains such as wheat, rice, barley, and in potato tubers decrease by 5–14% under elevated CO_2.

8.5.2 Environmental Factors on Plant Response to Elevated CO_2

The effects of elevated CO_2 on plants can vary depending on other environmental factors. While elevated CO_2 makes carbon more available, plants also require other sources including minerals obtained from the soil. Elevated CO_2 does not directly make these mineral elements more available and may even decrease the uptake of some elements. Another environmental factor that interacts with elevated CO_2 is atmospheric ozone (O_3), a gaseous toxin. Ground-level O_3 concentrations have been increasing worldwide due to increased emissions of pollutants that react to produce O_3. High atmospheric concentrations of ozone can cause damage to leaves and decreased plant growth and photosynthesis. The primary location of O_3 is in the internal tissue of leaves. Decreased openness of stomata under elevated CO_2 can therefore decrease exposure of sensitive tissue to ozone. Elevated CO_2 substantially decreases the negative effects of high ozone on photosynthesis, growth, and seed yield.

8.5.3 Differences among Plant Functional Types in Responses to Elevated CO_2

One of the most important determinations of species differences in response to elevated CO_2 is photosynthetic type. Most plant species (~90%) utilize a photosynthetic process known as C_3 photosynthesis. Other species use either of two physiologically distinct processes known as C_4 and Crassulacean acid metabolism photosynthesis. Figure 8.1 shows the pathways followed by C_3 plants and C_4 plants.

C_4 plants include most of the tropical and subtropical grassland and several other important crops such as maize, sugarcane, sorghum, and millets. C_4 plants use a biochemical pump to concentrate CO_2 at the locations within the leaf where the RUBISCO enzyme mediates incorporation of CO_2 by the Calvin-Benson photosynthetic cycle. Since CO_2 concentrations are already high within the bundle sheath cells, increasing atmospheric CO_2 concentrations above current levels has little direct effect on photosynthetic rates for C_4 species. C_4 species do respond to elevated CO_2 by decreasing stomatal conductance; this may lead to some indirect enhancement of photosynthesis by helping avoid water stress under drought conditions.

Stimulation of photosynthesis by elevated CO_2 in C_4 plants is only about one-third of that experienced by C_3 species. C_4 plants also show little or no enhancement of growth (dry matter production) in these studies (Ainsworth & Long 2005). The above figure depicts that C_4 plants are relatively unresponsive to elevation of atmospheric CO_2 above current ambient levels. In contrast to C_4 species, another group of plants, legumes, may be especially capable of responding to

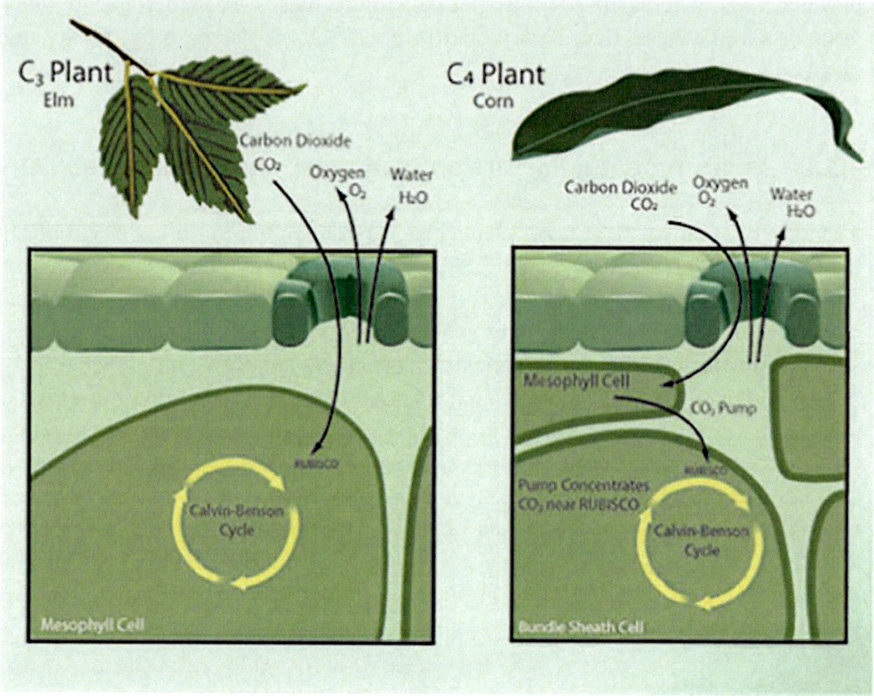

Fig. 8.1 C_3 pathway in Elm and C_4 pathway in corn

elevated CO_2 with increased photosynthesis and growth. For most plants, growth under elevated CO_2 can alter the internal balance between carbon (obtained in extra quantities through enhanced photosynthesis) and nitrogen (either unaffected or taken up in decreased amounts due to decreased uptake of water). Many studies in controlled environment showed that, compared to other plant species, legumes show greater enhancement of photosynthesis and growth by elevated CO_2. Decreases in tissue nitrogen concentrations under elevated CO_2 are also smaller for legumes than for other C_3 species.

8.5.4 Elevated CO_2 Effects on Root-to-Shoot N Signals

CO_2 enrichment influences root-to-shoot N signaling through its effects on xylem sap flow rate, NO_3^- assimilation, and root allocation. The availability of root-to-shoot N signals depends upon xylem sap flow for rapid signal delivery, and elevated CO_2 affects xylem flow rates. Elevated CO_2 decreases transpiration rates between 5 and 20% as stomata close in response to higher intercellular CO_2 concentration. Stomatal closure slows water uptake and thereby xylem sap flow rate. Decreased transpiration may impede mass flow of NO_3^- in the soil solution to roots, but this decrease may not slow delivery of N to shoots because N concentration in the xylem sap increases as xylem sap flow decreases, maintaining N delivery rates. Leaf N concentrations decline under prolonged growth at elevated CO_2. Photosynthetic acclimation can account for some of this decrease, but fertilization with NH_4NO_3 eliminates it, showing that increased N supply can compensate for the effects of elevated CO_2 through enhanced root N uptake and plant N assimilation. This suggests that elevated CO_2 interrupts root-to-shoot N signaling.

Amino acids in the phloem, potential signals of shoot N status, do not show a consistent response to elevated CO_2. By contrast, leaf and root auxins increase under-elevated CO_2 and stimulate root growth. However, other processes, such as carbohydrate transport or shoot NO_3^- assimilation, may limit the ability of increased root auxins to stimulate root N uptake.

In most tropical and subtropical plants and in temperate plants at high soil NO_3^-, most NO_3^- assimilation occurs in shoots because NO_3^- photo-assimilation in shoots is more energy efficient than respiratory-driven NO_3^- and NO_2^- reduction in roots. Elevated CO_2 inhibits shoot NO_3^- assimilation in C_3 plants, necessitating a greater reliance on root NO_3^- assimilation to maintain plant capacity for NO_3^- assimilation. In tobacco, 3 weeks of CO_2 enrichment enhances root NO_3^- assimilation and may compensate for decreasing shoot NO_3^- assimilation when there is sufficient root carbohydrate. A shift from shoot NO_3^- assimilation to root NO_3^- assimilation requires translocation of more carbohydrate to the roots to provide sufficient energy and carbon skeletons for these processes. NH_4^+ fertilization decreases the limitations of phloem carbohydrate transport on plant N status because NH_4^+ assimilation requires less carbohydrate.

8.5.5 Methodology for Studying Effect of CO_2

There are several good methods for studying the effect of CO_2 in plants such as eddy covariance and FACE, but FACE (Free-Air Carbon dioxide Enrichment) is a method used by ecologists and plant biologists that raises the concentration of CO_2 in a specific area and allows the response of plant growth to be measured. Experiments using FACE are required because most studies looking at the effect of elevated CO_2 concentrations have been conducted in labs and where there are many missing factors including plant competition. Measuring the effect of elevated CO_2 using FACE is a more natural way of estimating how plant growth will change in the future as the CO_2 concentration rises in the atmosphere. FACE also allows the effect of elevated CO_2 on plants that cannot be grown in small spaces (trees for example) to be measured. However, FACE experiments carry significantly higher costs relative to greenhouse experiment.

8.5.6 Change in Secondary Metabolites and Pest–Disease Reaction

Several environmental factors such as nutrient supply, temperature, light conditions, and atmospheric CO_2 concentrations can influence the levels of total phenolics and flavonoids in plants. As it is commonly known that secondary metabolites are associated with primary metabolism by the rates at which substrates are redirected from primary pathways to the secondary biosynthetic routes, several environmental factors affecting growth, photosynthesis, and other parts of primary metabolism will also influence secondary metabolism. One of these factors is CO_2, and the increases in atmospheric CO_2 due to climate change have a direct impact on plant secondary metabolites.

It has generally been assumed that increasing atmospheric CO_2 concentrations will increase plant carbon-based secondary or structural compounds (CBSSC) concentrations. These changes may have far-reaching consequences for herbivory and plant litter decomposition. Among such composition changes, decreased leaf N concentration has been widely reported. The concentration of carbon-based secondary or structural compounds (CBSSC) such as phenolics, terpenes, or structural polysaccharides (cellulose, hemicellulose, and pectin) is also expected to increase.

Recent experimental results provide evidence of increases in concentrations of soluble phenolics and condensed tannins but not in lignin, structural polysaccharides, or terpenes. They also show significant effects of these plant chemical changes on herbivores and little or any effects on decomposition. However, there is no consistent evidence of any of these effects at the complex long-term ecosystem level. The source-sink hypotheses (carbon nutrient balance hypothesis) and growth-differentiation balance hypothesis assume that elevated CO_2 concentration supports a comparative increase in carbon accessibility that is accumulated in total nonstructural carbohydrate (TNC) and carbon-based secondary metabolites (CBSM) when the provided carbon amounts exceed growth requirements. Under optimal CO_2 concentration conditions combined with nutrient resource restriction, which limit growths to a greater level than photosynthesis, plants showed an increase in the C/N ratios and excess of nonstructural carbohydrates. Factors such as high CO_2 concentration increase the carbon supply and consequently decreases the carbon demand, promoting a relative increase in carbon availability of carbon-based structural compounds that are components of the cell wall.

High atmospheric CO_2 concentrations often increase TNC concentrations in plants and possibly stimulate the secondary metabolism and antioxidant activity in plants. The studies evaluated the response of the tropical spider lily (*Hymenocallis littoralis*) to elevated levels of atmospheric CO_2 over four growing seasons indicated that a 75% increase in the ambient CO_2 concentration produced an 8% increase in pancratistatin, an 8% increase in *trans*-dihydronarciclasine, and a 28% increase in narciclasine, all of which are effective against lymphocytic leukemia and ovary sarcoma. In the early studies of Barbale and Madsen, a tripling of the atmospheric CO_2 concentration produced a modest (7%) increase in antioxidants in the leaves and fruits of tomato plants. Hence, it can be concluded that exposure of plant to high levels of CO_2 can increase the production of secondary metabolites and antioxidant.

Two hypotheses have been put forth to explain the increases of CBSM and antioxidants under high levels of CO_2. Bass considered that TNC in surplus of those required for protein production as the main factor affecting the increase of CBSM under elevated CO_2. On the other hand, scientists proposed the contribution of the amino acid phenylalanine as a common precursor for proteins and phenolics. The competition between proteins and phenolics for limiting phenylalanine would result in a swapping between protein and phenolics synthesis rates, reversing the link between protein and phenolics distribution. These two hypotheses reflect on the increase of CBSM under elevated CO_2 as a result of a metabolic excess of carbon with no physiological cost on growth. A negative relationship between CBSM and growth are anticipated because increases in growth along a gradient of increasing nutrient availability will result in less carbon being accessible for CBSM production.

8.5.7 Effect of Elevated CO_2 and Temperature on Crop–Disease Interactions

Elevated carbon dioxide concentrations and temperatures associated with climate change will have a substantial impact on plant–disease interactions. Changes in temperature affect both the host and the pathogen; thus, risk analyses must be conducted for each patho-system to determine the effects of climate change. Increased CO_2 levels can impact both the host and the pathogen in multiple ways. New races may evolve rapidly under elevated temperature and CO_2, as evolutionary forces act on massive pathogen populations boosted by a combination of increased fecundity and infection cycles under favorable microclimate within enlarged canopy. Studies have been performed under controlled conditions, and the effects of high CO_2 levels have been identified; however, field responses such as the adaptation of pathogens over time may be different. The climate influences the incidence as well as temporal and spatial distribution of plant diseases. The most likely effect of climate change in pole ward modifies agroclimatic zones, causing a shift in the geographical distribution of host pathogens. Elevated CO_2 concentration and temperature have an impact on plant–disease interaction, posing a higher threat perception of late blight (*Phytophthora infestans*) of potato and blast (*Magnaporthe grisea*) and sheath blight (*Rhizoctonia solani*) of rice. The effects of climate change on Phoma (*Leptosphaeria maculans*) in rape seed were observed using a model in combination with climate change that predicts temperature and rainfall under CO_2 emission scenarios for the 2020 and 2050s in the United Kingdom, as well as sporulation of teleomorphs on climate change. The major predicted results of climate change-related increases in temperature, moisture, and CO_2 can impact all three legs of the plant disease triangle in various ways. Precisely predicting the impact of climate change on plant disease is a tricky business. Other study indicated that colony area, number of hyphal tips, conidiophores, and conidia per colony were significantly greater at temperature ranging from 22 to 28 °C compared to standard conditions. Disease indices (0–7) also varied in different trials due to the quantity of inoculums, but were never influenced only by increased CO_2. On the contrary, elevated temperature and CO_2 significantly stimulated disease index. There was a clear increment of growth of the pathogen, fecundity, and severity of the disease observed at the higher temperature–higher CO_2 combination as compared to control temperature combinations, particularly with higher CO_2 at standard temperatures.

8.6 Ozone Layer

The ozone layer is not really a layer at all, but has become known as such because most ozone particles are scattered between 19 and 30 km (12–30 miles) up in the earth's atmosphere in a region called the stratosphere. The concentration of ozone in the ozone layer is usually less than 10 parts ozone per million. Ozone is created in the stratosphere when UV radiation from the sun strikes molecules of oxygen (O_2) and causes the two oxygen atoms to split apart. If a freed atom bumps into another O_2, it joins up, forming ozone (O_3). This process is known as photolysis.

$$O_2 + UV - C \rightarrow O + O$$

$$O + O_2 \rightarrow O_3 + \text{heat}$$

8.6.1 Ozone Layer Depletion

Ozone layer depletion is the wearing out or reduction of the amount of ozone in the stratosphere. Ozone depletion occurs when chlorofluorocarbons (CFCs) and halons-gases found in aerosol spray cans and refrigerants-are released into the atmosphere. CFCs and halons cause chemical reactions that break down ozone molecules, reducing ozone's UV radiation-absorbing capacity.

8.6.2 Causes of Ozone Depletion

Ozone depletion occurs when the natural balance between the production and destruction of stratospheric ozone is tipped in favor of destruction. Although natural phenomena can cause temporary ozone loss, chlorine and bromine released from man-made compounds such as CFCs are now accepted as the main cause of this depletion. The groups of compounds known as the chlorofluorocarbons (CFCs) were most likely the main source of ozone depletion. However, this idea was not taken seriously until the discovery of the ozone hole over Antarctica in 1985 by the survey. Chlorofluorocarbons are not "washed" back to the earth by rain or destroyed in reactions with other chemicals. They simply do not break down in the lower atmosphere, and they can remain in the atmosphere from 20 to 120 years or more. As a consequence of their relative stability, CFCs are instead transported into the stratosphere where they are eventually broken down by ultraviolet (UV) rays from the sun, releasing free chlorine. The chlorine becomes actively involved in the process of destruction of ozone. The net result is that two molecules of ozone are replaced by three of molecular oxygen, leaving the chlorine free to repeat the process:

$$Cl + O_3 = ClO + O_2$$

$$ClO + O = Cl + O$$

Ozone is converted to oxygen, leaving the chlorine atom free to repeat the process up to 100,000 times, resulting in a reduced level of ozone.

8.6.3 Ozone Hole

The ozone hole is not technically a "hole" where no ozone exists, but rather a region in the stratosphere, particularly over the Antarctic region, where exceptionally depleted ozone exists. Technically, the term "ozone hole" should be applied to regions where stratospheric ozone depletion is so severe that levels fall below 200 Dobson Units (D.U.), the standard measure of stratospheric ozone. The normal ozone concentration is about 300–350 D.U. Such ozone loss now occurs every spring time above Antarctica, and to a lesser extent the Arctic, where special meteorological conditions and very low air temperatures accelerate and enhance the destruction of ozone loss caused by man-made ozone depleting chemicals (ODCs).

8.6.4 Ozone Depletion over India

According to National Ozone Centre, New Delhi, there is no trend to show total ozone depletion over India. In addition, the Indian Meteorological Department also points out that for the period 1956–1986, "ozone measurements exhibit year to year variability, but do not show any increasing or decreasing trend over India." The National Physical Laboratory, which has been working on incident UV radiation levels, claims that during the summer, at noon, UV-B radiation with a wavelength of 290 nanometer (nm) is equivalent to levels attained in the Antarctica during the ozone hole period. They warn that even a slight depletion of the ozone layer over India may lead to large percentage changes in UV-B radiation over the country.

8.6.5 Effect of Ozone Layer Depletion in Plants

It is a known fact that physiological and development processes of plants are affected by UV-B radiation. Scientists believe that an increase in UV-B level would necessitate using more UV-B tolerant cultivars and breeding new tolerant ones in agriculture.

In forests and grasslands, increased UV-B radiation due to ozone depletion is likely to cause change in species composition (mutation), thus altering the biodiversity in different ecosystems. UV-B radiation could also affect the plant community, indirectly resulting in changes in plant form, 2° metabolism, and so on. These changes can have important implications for plant competitive balance, plant pathogens, and biogeochemical cycles.

8.6.6 Some Other Effects of Ozone Depletion on Plants

8.6.6.1 Direct Effect

(a) DNA damage: Cyclo-butane dimer formation.
(b) Photosynthesis: Disturbance for photosynthesis thylakoid membrane disfunctioning stomatal functioning altered.
(c) Membrane functioning: Peroxidation of unsaturated fatty acids, damage to membrane proteins.

8.6.6.2 Indirect Effects

(a) Plant morphogenic effect: Leaf thickness, leaf angle, plant architecture.
(b) Plant morphology: Senescence, flowering, reproduction.
(c) Chemical composition of plant: Tannin, lignin, flavonoids.

8.6.7 Tolerance in Plants against Ozone Depletion or UV Radiation

(a) The reproductive parts of plants, such as pollen and ovules, are well shielded from solar UV-B radiation.
(b) Only a small proportion of UV-B radiation striking the leaf penetrates into the inner tissues.
(c) Increased thickness of leaves reduces the proportion of inner tissues exposed to UV-B radiation.
(d) Have repair mechanism in plants (include repair systemic for DNA damage).

8.6.8 Mechanism of UV-B Penetration of Plant Tissue

In order for you to have any direct or developmental effect on photosynthetic productivity, UV-B must penetrate through the leaf and be absorbed by chromophores associated with the photosynthetic apparatus or associated genes or gene products. Leaves absorb over 90% of incident UV-B; leaf surface reflectance in this wavelength range is generally below 10% and this is negligible transmission of UV-B through leaves. Cellular compounds that can absorb UV-B directly include nucleic acid, protein, and quinines. The leaves also contain water-soluble phenolic pigments such as flavonoids, which absorb UV-B radiation, but not photosynthetically active radiation. These UV-B absorbing pigments are present throughout the leaf but accumulate significantly in adaxial epidermal cells.

8.6.9 Plants and UV Radiation

Plants, due to their nonmotile nature, generally have a higher rate of UV tolerance than animals. Plant secondary metabolites aid in defense against both abiotic and biotic stress factors. Plants are capable of absorbing or reflecting harmful UV rays via thick layers of waxy cutin or submurin on the cell walls and intracellular accumulation of chemical substances such as flavanols or phenolics. The biological effects of UV radiation on plants include altered growth responses, reproductive deformities, epigenetic variations, plant susceptibility to biotic factors, premature senescence, damage to the photosynthetic apparatus, and altered conformation of membrane structures.

8.6.10 UV-Induced DNA Damage

UV-C/-B radiation is directly absorbed by DNA, inducing lesions that inhibit vital cellular functions such as transcription and DNA replication. UV-A is comparatively less efficient in lesion induction but triggers the production of reactive oxygen species (ROS).

The primary UV-induced DNA lesions include cyclobutane pyrimidine dimers and secondary lesions, 6-4 pyrimidine–pyrimidone photoproducts (6-4 PP). In order to respond to this damage, plants employ specific mechanisms. In light conditions, photoreactivation catalyzes dimer monomerizations, while during dark conditions, nucleotide excision repair (NER) excises these helix-distorting lesions. Finally, residual lesions are bypassed through trans-lesion synthesis (TLS).

8.7 UV-B Avoidance and Defense Mechanism

Plants use certain signaling mechanisms to enhance avoidance at morphological, physiological, biochemical, and molecular levels. They show adaptation to environmental stresses, sometimes referred to as "Plant Memory." There is growing evidence that plants memorize exposure to biotic and abiotic stresses through epigenetic mechanisms at the cellular level. UV-B radiation is a key environmental signal that is specifically perceived by plants to promote UV acclimation and survival in sunlight. Scattering and reflection of UV-B radiation is achieved by epidermal and cuticular structures and other leaf optical properties such as waxy layer, leaf hairs, and leaf bladders. Figure 8.2 shows the detail of mechanism of DNA damage and repair.

8.7.1 Avoidance at Morphological Level

Leaf morphological and anatomical changes, such as increased epicuticular wax content, increase in cuticle thickness, wider epidermis and palisade layers, are some of the modifications of plants employ to avoid UV-B exposure. In general, it

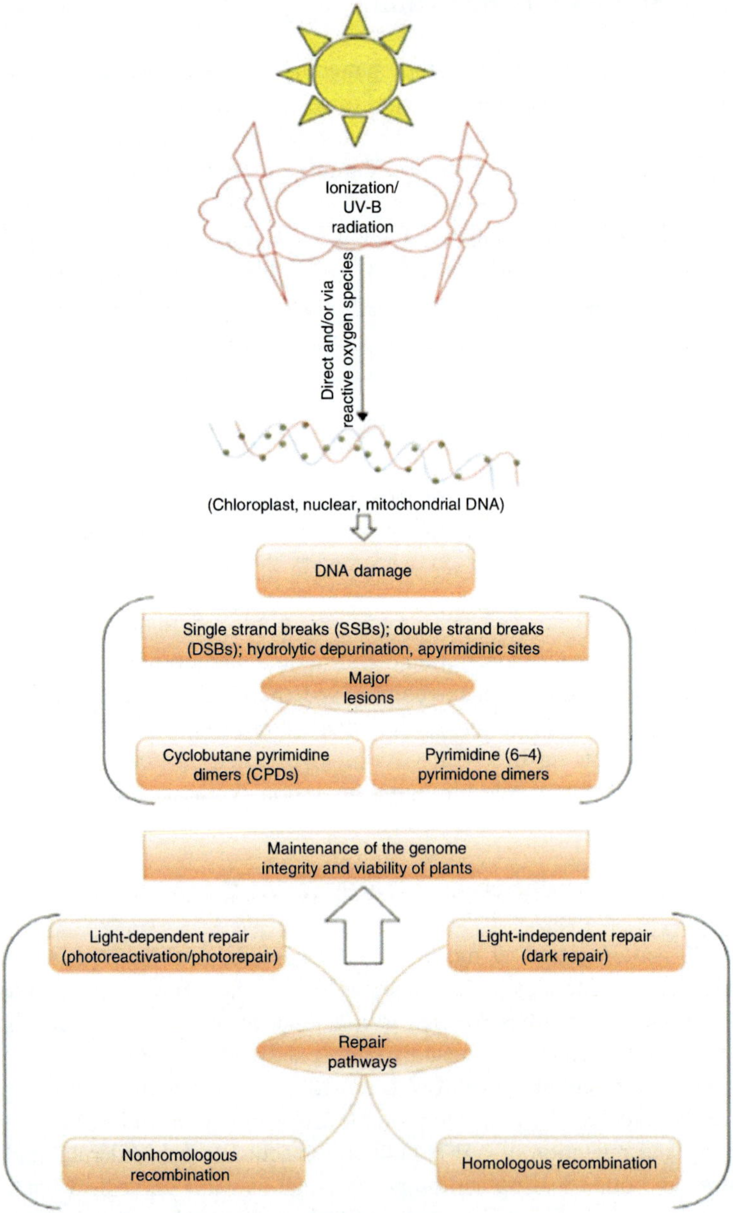

Fig. 8.2 Mechanism of DNA damage and repair

possesses a suit of mechanisms that act either to prevent the absorption of harmful and excess radiation or to mitigate against the damage that such radiation can cause once it is absorbed, as shown in Fig. 8.3. Increased wax provides protection by reflecting 10–30% of the incident UV-B radiation in many plants.

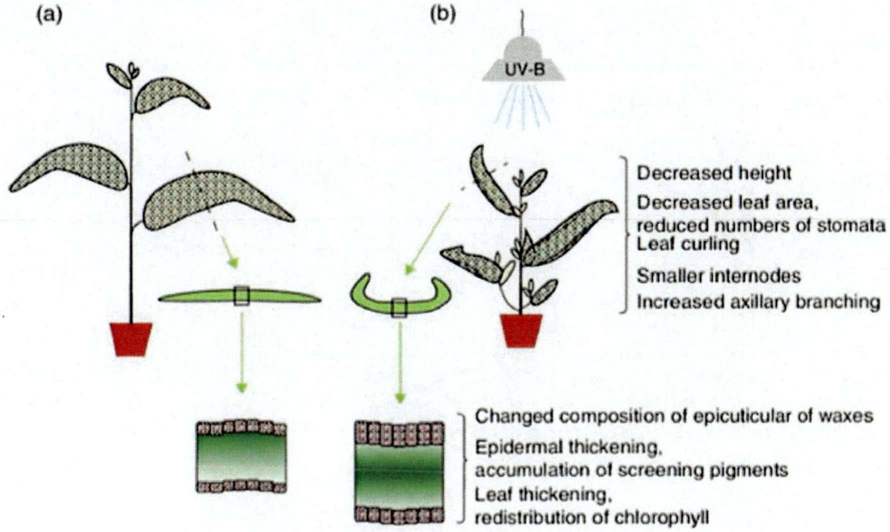

Fig. 8.3 Preventive mechanism at morphological level

8.7.2 Avoidance at Biochemical Level

8.7.2.1 Possible Role of Pectin Endocytosis in UV-B Avoidance

Poly-galacturonic acid and galacturonic acid are reported to generate superoxide upon UV-B irradiation in the presence of hydrogen peroxide. In root cells, pectin is a major component of the poly-galacturonic acid cell wall. Pectin, cross-linked with boron and calcium, has been shown to be internalized and transported by endocytosis. For root tropisms, endocytic vesicle recycling is a critical process and is especially active in the root apex transition zone, which is located in between the apical meristem and basal elongation zone. It is reasonable to expect the importance of cross-linked cell wall pectin for ROS homeostasis, because UV-B easily breaks down cell wall components and this might act as a cue for the ascorbate genesis at irradiated organ sides in order to recover the cellular redox balance. L-ascorbic acid is a well-known reducing agent, controlling cellular redox homeostasis via glutathione-ascorbate cycle. The whole mechanism can be easily understood from Fig. 8.4. Of course, internalized cell wall pectin within endocytic vesicles and endosomes might also be a potential source for the L-ascorbate synthesis.

8.7.3 Avoidance at the Molecular Level

(a) **DNA repair**: The repair of DNA is essential for the survival of organisms; otherwise, genomic integrity will not be maintained.
    DNA repair is performed in two conditions:

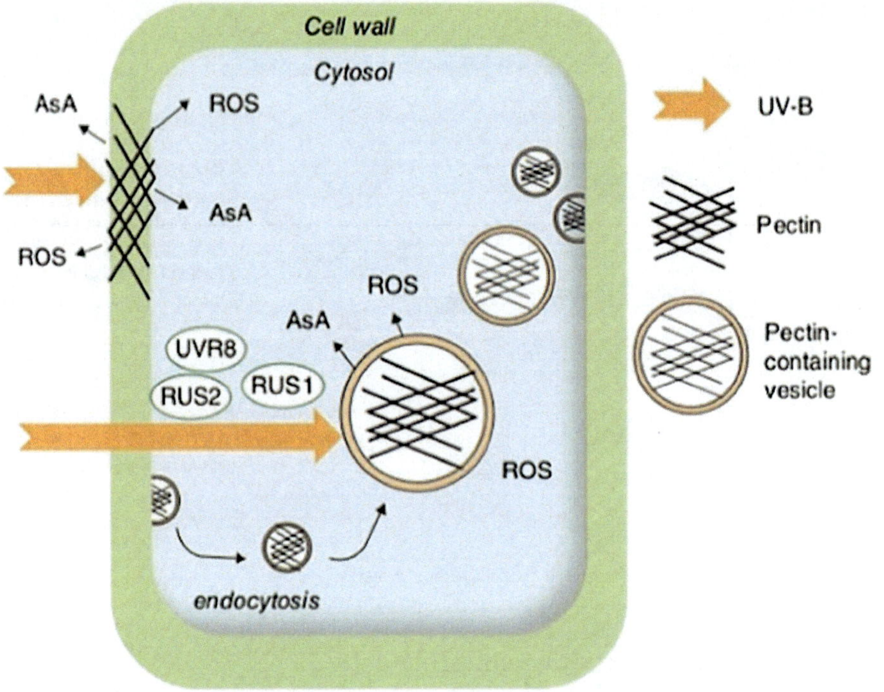

Fig. 8.4 Pectin as a redox regulator during UV-B response in cells of the root apex transition zone

- In light conditions, photoreactivation catalyzes dimer monomerizations.
- During dark conditions, nucleotide excision repair (NER) excises helix-distorting lesions and base excision repair (BER) repairs oxidized or hydrated bases.

(b) **Genes and avoidance**: UV-B specific signaling components orchestrate plant protection against UV-B irradiation. UV Resistance Locus 8 (UVR8) is a key photoreceptor of the photomorphogenic pathway, regulating a range of genes with important roles in UV protection and the repair of UV damage. It was observed the association of UVR8 chromatin with the promoter region of Elongated Hypocotyl 5 (HY5).

(c) **UV-B perceived by UVR8 strongly inhibits shade avoidance**: In plants, UV-B is perceived by photoreceptor protein UVR8. Along with regulating photo-protective responses, UV-B dramatically inhibits stem elongation. When grown in dense stands, plants use reflected far-red light signals from neighbors to detect the threat of shading.

UV-B perceived by UVR8 provides a potent sunlight signal that inhibits shade avoidance. UVR8 activation stimulates multiple pathways that converge to block biosynthesis of the plant growth hormone. After the emergence of plants and growing in sunlight, long-term acclimation is the most important response for coping with UV exposure. Thus, patterns of gene expression with relevance for long-term acclimation of plants to solar UV radiation were determined in the presence of UV-A and high photosynthetically active radiation (PAR).

8.8 UV-B and Secondary Metabolites

Another adaptive mechanism against enhanced UV-B radiation is increased production of secondary metabolites in leaf tissues. Some studies indicated that UV-B-absorbing compounds increase from 10 to 30% in agronomic crops. One of the most effective defensive mechanisms against UV-B radiation in higher plants is the accumulation of a diverse range of phenolic metabolites. UV-B radiation transmitted after reflection by the epicuticular wax layer reaches the epidermal layer. This layer is known to accumulate most of the secondary metabolites, such as phenolics and flavonoids, which absorb or screen UV-B radiation and shield the underlying tissues against harmful UV-B radiation. The accumulation of leaf condensed tannins, salicylates, and flavonoids was increased more in females than in males by elevated UV-B and facilitated by heat.

Bibliography

Abbas T, Dutta A (2011) CRL4cdt 2: master coordinator of cell cycle progression & genome stability. Cell Cycle 10(2):241–249

Abrol YP, Bagga AK, Chakravorty NVK, Wattal PK (1991) Impact of rise in temperature on the productivity of wheat in India. In: Abrol YP et al (eds) Impact of global climate changes in photosynthesis and plant productivity. New Delhi, pp 787–789

Aggarwal PK (2000) Climate change and agriculture: information needs and research priorities. Indian Agricultural Research Institute, New Delhi

Aggarwal PK (2003) Impact of climate change on Indian agriculture. J Plant Biol 30(2):189–198

Aggarwal PK, Kalra N (1994) Simulating the effect of climatic factors, genotype and management on productivity of wheat in India. Indian Agricultural Research Institute Publication, New Delhi, p 156

Aggarwal PK, Mall RK (2002) Climate change and rice yields in diverse agro-environments of India. II. Effect of uncertainties in scenarios and crop models on impact assessment. Clim Chang 52(3):331–343

Aggarwal PK, Sinha SK (1993) Effect of probable increase in carbon dioxide and temperature on productivity of wheat in India. J Agric Meteorol 48(5):811–814

Aggarwal PK, Kumar SN, Pathak H (2010) Impacts of climate change on growth and yield of rice and wheat in the Upper Ganga Basin. Indian Agricultural Research Institute (IARI)

Allen LH Jr (n.d.) Plant response to rising carbon dioxide and potential interactions with air pollutants. J Environ Qual 19:15–34

Allen, R. G., Smith, M., Pereira, L. S. and Perrier, A. 1994. An Update for the calculation of reference evapotranspiration. ICID Bull, 43(2): 35–92

Allen RG, Smith M, Pereira LS, Raes D, Smith M (1998) Crop evapotranspiration. Guidelines for computing crop water requirements. In: FAO irrigation and drainage. FAO, Rome, p 300

Arora M, Goel NK, Singh P (2009) Evaluation of temperature trends over India. Hydrol Sci J 50:93

Attri SD, Rathore LS (2003) Simulation of impact of projected climate change on wheat in India. Int J Climatol 23:693–705

Baas WJ (1989) Secondary plant compounds, their ecological significance and consequences for the carbon budget. Introduction of the carbon/nutrient cycle theory. In: Lambers H, Cambridge M, Konings H, Pons TL (eds) Causes and consequences of variation in growth rate and productivity of higher plant. SPB Academic Publishing, The Hague, pp 313–340

Bachelete D, Kropff MJ (1995) The impact of climatic change on agro climatic zones in Asia. In: Matthews RB et al (eds) Modeling the impact of climate change on rice production in Asia. CAB International, UK

Baker EA (1982) Chemistry and morphology of plant epicuticular waxes. In: Cutler DF, Alvin KL, Price CE (eds) The plant cuticle. Academic Press, London, pp 139–166

Baker JT, Allen LH Jr, Boote KJ, Jones P, Jones JW (1989) Response of soybean to air temperature and carbon dioxide concentration. Crop Sci 29:98–105

Ballaré CL (2003) Stress under the sun: spot light on UV-B responses. Plant Physiol 132:1725–1727

Barbale D (1970) The influence of the carbon dioxide on the yield and quality of cucumber and tomato in the covered areas. Augsne un Raza (Riga) 16:66–73

Booker FL (2000) Influence of carbon dioxide enrichment, ozone and nitrogen fertilization on cotton (*Gossypium hirsutum* L.) leaf and root composition. Plant Cell Environ 23:573–583

Britt AB (1999) Molecular genetics of DNA repair in higher plants. Trends Plant Sci 4(1):20–25

Brown S, Lincke D, Nicholls RJ, Hinkel J (2015) The impacts of sea-level rise on European coasts in a 2°C world. Results and analysis of task 6.5 prepared as part of IMPACT2C: quantifying project impacts under 2 °C warming. University of Southampton, Southampton, GB

Bryant JP, Chapin FS, Klein DR (1983) Carbon nutrient balance of boreal plants in relation to vertebrate herbivory. Oikos 40:357–368

Chala D, Meseret T (2018) Effect of elevated CO_2 and temperature on crop-disease interactions under rapid climate change. Int J Environ Sci Nat Res 13(1):555851

Chand R, Singh C (2015) Movements of western disturbance and associated cloud convection. J Indian Geophy Union 19:62–70

Ciais P, Reichstein M, Viovy N, Granier A, Ogee J et al (2005) Europewide reduction in primary productivity caused by the heat and drought in 2003. Nature 437:529–533

Cicerone RJ, Nurse P (2014) Climate change: evidence and causes. The Royal Society

Cline RW (1999) Global warming and agriculture: impact estimation by country. Center for Global Development and the Peterson Institute for International Economics, Washington, DC

Cline W (2008) Global warming and agriculture: impact estimates by country. Peterson Institute for International Economics, Washington, DC

Cure JD, Acock B (1986) Crop responses to carbon dioxide doubling: a literature survey. Agric For Meteorol 38:127–145

Dash SK, Hunt JCR (2007) Variability of climate change in India. Curr Sci 93

De US, Mukhopadhyay RK (1998) Severe heat wave over the Indian subcontinent in 1998, in perspective of global climate. Curr Sci 75(12):1308–1311

Dixon GR, Collier RH, Bhattacharya I (2014) An assessment of the effects of climate change on horticulture. In: Dixon GR, Aldous DE (eds) Horticulture: plants for people and places, vol 2. Springer, Dordrecht, pp 817–857

DOD (2002) Annual Report 2001/02. Department of Ocean Development, New Delhi

Easlon HM, Bloom AJ (2013) The effects of rising atmospheric carbon dioxide on shoot–root nitrogen and water signaling. Front Plant Sci 4:304

Felgines C, Texier O, Morand C, Manach C, Scalbert A, Regerat F, Remesy C (2000) Bioavailability of the flavone naringenin and its glycosides in rats. Amer J Physiol Gastrointest Liver Physiol 279:1148–1154

Fine PVA, Miller ZJ, Mesones I, Irazuzta S, Appel HM, Stevens MHH (2006) The growth defense tradeoff and habitat specialization by plants in Amazonian forest. Ecology 87:150–162

Fischer G, Frohberg K, Parry ML, Rosenzweig C (1994) Climate change and world food supply, demand and trade. Who benefits, who loses? Glob Environ Chang 4(1):7–23

Francis M (1999) Simulating the impact of increase in temperature and CO2 on growth and yield of rice, M.Sc. Thesis (unpublished), Indian Agricultural Research Institute, New Delhi

Garrett KA, Nita M, De Wolf ED, Esker PD, Gomez-Montano L 2015 Plant pathogens as indicators of climate change. Climate Change pp. 325-338

Geethalakshmi V, Lakshmanan A, Rajalakshmi D, Jagannathan R, Sridhar G et al (2011) Climate change impact assessment and adaptation strategies to sustain rice production in Cauvery basin of Tamil Nadu. Curr Sci 101:342–347

Glenn DM, Kim SH, Ramirez-Villegas J, Läderach P (2013) Response of perennial horticultural crops to climate change. In: Janick J (ed) Horticultural reviews, vol 41, 1st edn. John Wiley & Sons, Hoboken, NJ, pp 47–129

Herms DA, Matson WJ (1992) The dilemma of plants: to grow or defend. Q Rev Biol 67:283–335

Hogy P, Fangmeier A (2009) Atmospheric CO_2 enrichment affects potatoes: tuber quality traits. Eur J Agron 30:85–94

Huber DG, Gulledge J (2011) Extreme weather and climate change: understanding the link and managing the risk. Pew Center for Climate and Energy Solutions, Arlington

Ibrahim MH, Jaafar HZE, Rahmat A, Abdul RZ (2011) Effects of nitrogen fertilization on synthesis of primary and secondary metabolites in three varieties of kacip fatimah (*Labisia pumila* Blume). Int J Mol Sci 12:5238–5254

Idso SB, Kimball BA, Pettit GR III, Garner LC, Pettit GR, Backhaus RA (2000) Effects of atmospheric CO_2 enrichment on the growth and development of *Hymenocallis littoralis* (Amaryllidaceae) and the concentrations of several antineoplastic and antiviral constituents of its bulbs. Am J Bot 87:769–773

India Meteorological Department (IMD) (1995) Report on cyclonic disturbances over North Indian Ocean during 1995. Regional Specialized Meteorological Centre for Tropical Cyclones Over North Indian Ocean (Abridged report for circulation during the annual meeting of WMO/ESCAP Panel on Tropical Cyclones)

Intergovernmental Panel on Climate Change (IPCC) (2001) Climate change: synthesis report 2001. Summary for Policy makers, Wembley

Intergovernmental Panel on Climate Change (IPCC) (2007) Climate change 2007: impacts, adaptation and vulnerability. In: Contribution of Working Group II to the Fourth Assessment Report of the Intergovernmental Panel on Climate Change. Cambridge University Press, Cambridge

Intergovernmental Panel on Climate Change (IPCC) (2014) Climate change 2014: synthesis report. In: Contribution of Working Group I, II and III to the FIIH Assessment Report of the Intergovernmental Panel on Climate Change. IPCC, Geneva

Izaurrale RC, Rosenberg NJ, Brown RA, Homson AM (2003) Integrated assessment of Hardley Center (HadCM2) climate change impacts on agricultural productivity and irrigation water supply in the conterminous United States Part II. Regional agricultural production in 2030 and 2095. Agr Forest Meterol 117:97–122

Jones CG, Hartley SE (1999) A protein competition model of phenolic allocation. Oikos 86:27–44

Kang Y, Khan S, Ma X (2009) Climate change impacts on crop yield, crop water productivity and food security – a review. Prog Nat Sci 19:1665–1674

Kemp AC, Horton BP, Donnelly JP, Mann ME, Vermeer M et al (2011) Climate related sea-level variations over the past two millennia. Proc Natl Acad Sci 108:11017

Khan SA, Kumar S, Hussain MZ, Kalra N (2009) Climate change, climate variability and Indian agriculture: impacts vulnerability and adaptation strategies. In: Climate Change Crops. Springer, Berlin, pp 19–38

Kitoh A, Endo H (2016) Changes in precipitation extremes projected by a 20-km mesh global atmospheric model. Weather Climate Extremes 11:41–52

Kundzewicz ZW, Matczak P (2015) Hydrological extremes and security. Proc IAHS 366:44

Lakshmikumar TV, Barbosa H, Rao KK, Jothi EP (2012) Some studies on the frequency of extreme weather events over India. J Agricul Sci Tech 14:1343–1356

Lambers H (1993) Rising CO_2, secondary plant metabolism, plant herbivore interactions and litter decomposition. Theoretical considerations. Vegetatio 104(105):263–271

Leakey ADB, Ainsworth EA, Bernacchi CJ, Rogers A, Long SP, Ort DR (2009) Elevated CO_2 effects on plant carbon, nitrogen, and water relations: six important lessons from FACE. J Exp Bot 60:2859–2876

Luo Y, Field CB, Mooney HA (1994) Predicting responses of photosynthesis and root fraction to elevated CO2: interactions among carbon, nitrogen and growth. Plant Cell Environ 17:1195–1204

Madsen E (1971) The influence of CO_2-concentration on the content of ascorbic acid in tomato leaves. Ugeskr Agron 116:592–594

McDonald EP, Erickson JE, Kruger EL (2002) Can decreased transpiration limit plant nitrogen acquisition in elevated CO_2. Funct Plant Biol 29:1115–1120

Mcelrone AJ, Reid CD, Hoye KA, Hart E, Jackson RB (2005) Elevated Cd reduces disease incidence and severity of a red maple fungal pathogen via changes in host physiology and leaf chemistry. Glob Chang Biol 11(10):1828–1836

McGuire AD, Melillo JM, Joyce LA (1995) The role of nitrogen in the response of forest net primary production to elevated atmospheric carbon dioxide. Annu Rev Ecol Syst 26:473–503

Melloy P, Hollaway G, Luck JO, Norton ROB, Aitken E et al (2010) Production and fitness of Fusarium pseudograminearum inoculum at elevated carbon dioxide in FACE. Glob Chang Biol 16(12):3363–3373

Milus EA, Seyran E, McNew R (2006) Aggressiveness of *Puccinia striiformis* f. sp. *tritici* isolates in the south-central United States. Plant Dis 90(7):847–852

Neenu S, Biswas AK, Rao AS (2013) Impact of climatic factors on crop production-a review. Agric Rev 34:97–106

Niranjankumar K, Rajeevan M, Pai DS, Srivastava AK, Preethi B (2013) On the observed variability of monsoon droughts over India. Weather and Climate Extremes 1:42–50

Panuelas J, Estiarte M (1998) Can elevated CO_2 affect secondary metabolism and ecosystem function? Trends Ecol Evol 13:20–24

Paroda RS, Praduman K (2000) Food production and demand in South Asia. Agric Econ Res Rev 13:1–24

Pathak H, Byjesh K, Chakrabarti B, Aggarwal PK (2011) Potential and cost of carbon sequestration in Indian agriculture: estimates from long-term field experiments. Field Crop Res 120:102–111

Peng S, Huang J, Sheehy J, Laza R, Visperas R et al (2004) Rice yields decline with higher night temperature from global warming. P Natl Acad Sci USA 101:9971–9975

Pettit GR, Gaddamidi V, Herald DL, Singh SB, Cragg GM, Schmidt JM, Boettner FE, Williams M, Sagawa Y (1986) Antineoplastic agents, 120. Pancratium littorale. J Nat Prod 49:995–1002

Porter JR, Semenov MA (2005) Crop responses to climatic variation. Phil Trans R Soc A 360: 2021–2035

Pritchard S, Rogers H, Prior SA, Peterson C (1999) Elevated CO_2 and plant structure: a review. Glob Chang Biol 5(7):807–837

Reichardt PB, Chapin FS, Bryant JP, Mattes BR, Clausen TP (1991) Carbon nutrient balance as a predictor of plant defense in Alaskan balsam poplar: a potential importance of metabolite turnover. Oecologica 88:401–406

Rosenzweig C, Tubiello FN, Goldberg R, Mills E, Bloomfield J (2002) Increased crop damage in the U.S. from excess precipitation under climate change. Glob Environ Chang 12:197–202

Saravanakumar V (2015) Impact of climate change on yield of major food crops in Tamil Nadu, India. SANDEE 91:91–15

Schnepf R, Chite RM (2005) U.S. agriculture after Hurricane Katrina: status and issues. The Library of Congress, Congressional Research Service

Schulze ED, Bloom AJ (1984) Relationship between mineral nitrogen influx and transpiration in radish and tomato. Plant Physiol 76:827–828

Shaner D, Boyer JS (1975) Relationship between nitrate supplied by transpiration stream and nitrate reductase-activity in maize leaves. Plant Physiol 56:61–61

Strain BR and Bazzaz FA (1983) CO_2 and plants. The response of plants to rising levels of atmospheric carbon dioxide, In American Association for the Advancement of the Science. Selected symposium 84 Lemon, E.R. 177–282, West view Press

Taub D (2010) Effects of rising atmospheric concentrations of carbon dioxide on plants. Nature Education Knowledge 3(10):21

Tester RF, Morrison WR, Ellis RH, Piggott JR, Batts GR et al (1995) Effects of elevated growth temperature and carbon dioxide levels on some physicochemical properties of wheat starch. J Cereal Sci 22:63–71

Thakur P, Kumar S, Malik JA, Berger JD, Nayyar H (2010) Cold stress effects on reproductive development in grain crops: an overview. Environ Exp Bot 67:429–443

The Directorate of Economic and Statistics in the Ministry of Agriculture, Government of India (2014) Agricultural statistics at glance

The Planning Commission, Government of India (1991–92) Economy and the plan: an overview

Udmale PD, Ichikawa Y, Kiem AS, Panda SN (2014) Drought impacts and adaptation strategies for agriculture and rural livelihood in the Maharashtra state of India. Open Agric J 8:41–47

Van der Velde M, Tubiello FN, Vrieling A, Bouraoui F (2012) Impacts of extreme weather on wheat and maize in France: evaluating regional crop simulations against observed data. Clim Chang 113:751–765

Williams M, Shewry PR, Lawlor DW, Harwood JL (1995) The effects of elevated temperature and atmospheric carbon dioxide concentration on the quality of grain lipids in wheat (Triticum aestivum L.) grown at two levels of nitrogen application. Plant Cell Environ 18:999–1009

You L, Mark RW, Fang C, Wood S (2005) Impact of global warming on Chinese wheat productivity. Environment and Production Technology Division. International Food Policy Research Institute

Climate Change Effect on Fruit Production

9

Abstract

Climate change has a profound impact on a crop's optimal growth period. Genetically, crop production is sensitive to climate variability. Climate change alters the photoperiod and temperature required for plant growth and agricultural systems. Climate change also affects the pollination process of the plant and renders the flowers less viable and more sterile. In this chapter, all the details of crop modifications and crop divergence have been incorporated. The mitigation strategies and prospects for genetic manipulation of crops to maximize production in the future have also been studied.

Keywords

Climate change · Stress · Pollination · Genetic engineering · RuBisCO

Climate change is a continuous process and it has become recognizable in the agricultural field from the past few years when it has started a significant and lasting effect on crop production. The reasons for climate change are not completely known today, but as per the available information, anthropogenic activities such as industrialization and mechanization may contribute up to some extent. The effects of temperature generated by global warming on crop plants are the major among all the climate change effects. It is again responsible for other stresses such as drought or moisture stress, salinity, and floods and water logging in coastal areas due to melting of polar ice and increased sea level. As carbon dioxide increases, the climate warms, with potentially disastrous outcomes, and it has become necessary to build and maintain a sustainable society and to preserve the productive natural systems that support us. Renewable resources such as forests, biodiversity, soils, and fresh water must be treated in ways that ensure their long-term productivity. Their capacity for renewal must be understood and respected. Therefore, it is clearly understood that

L. Ahmad et al., *Climate Change and Agrometeorology*,
https://doi.org/10.1007/978-981-99-4863-5_9

climate change will have a greater impact on global food security in the near future, and the limited availability of literature indicates that research on shifting cropping seasons, growth and yield patterns, and pest and disease scenarios under changing climatic conditions is needed. At the same time, producers and processors are already or will be faced with climate-driven problems with almost no knowledge or experience to develop effective management strategies. Consequently, future management decisions will be based on a higher level of uncertainty, which is unavoidable in future climate impacts predictions. Managing and communicating this uncertainty will be a challenge for researchers. Additional research will be needed to provide data-driven information for growers, processors, governments, and policy decisions about horticulture crops and the use of traditional knowledge, which has been lacking.

Climate change on fruit cultivation:

- Climate change sends Indian apple farmers up the Himalayas
- Apples in the Himalayan foothills are facing the worst effects of climate change
- Erratic rains dilute sugar content of the crop

Temperate fruit-growing regions in India:

- North-western Himalaya (Jammu & Kashmir, Himachal Pradesh, Uttarakhand)
- Parts of north-eastern Himalayas, Andhra Pradesh, Sikkim, Meghalaya, Darjeeling

Features of temperate region:

- Extreme cool winters and mild summers
- Growing period ranges from greater than 90 to 120 days
- Fragile ecosystem with harsh and difficult terrain
- Undulating sloppy terraced land, high soil erosion, and low depths
- Chilling requirement for temperate fruits is mandatory

Impact of climate change in Himalayan states (erratic climate/change climate):

- Variation in temperate rise is visible
- Early melting of glaciers and floods
- Change in cropping pattern/system
- Warming and extending winter's erratic changes
- Erratic and reduced winter precipitation and snow fall
- Depletion of ground water scarcity
- Shift in apple cultivation toward higher altitudes and cold arid areas
- Shift in ecological zones

- Majority of the apple orchards have become unproductive over the year due to low chilling and incidence and resurgence of insects and diseases of horticultural crops
- Mechanical damage during active growing stage directly responsible for canker and gummosis
- Hailstorms and low temperature during spring is a common phenomenon now
- Crop failure in high chill fruit crops under low altitudes
- Dry spell during late spring and early summer

9.1 Climate Change on Fruit Crops

The fruit crops suffer greater losses due to erratic weather conditions and overall changing climate. The early blooms followed by late snowfall cause severe damage to different fruit crops. The late snowfall proves to be devastating for the blossoms and it can be seen in Fig. 9.1.

9.1.1 Effect on Flowering

- Bud burst is advanced and the onset of growth occurs earlier as in apple where most of the trees sprout 2–3 weeks earlier than normal sprouting in mid-April.

Fig. 9.1 Damage caused by late snowfall on blossoms

- Due to change in bud opening in March, they may become susceptible to frost damage.

9.1.2 Effect on Fruit Quality

- Apples fail to attain their characteristic deep red color if night temperature is high.
- High temperature and moist stress increases sunburn and cracking in apple, apricot, cherry, pomegranate, and so on.
- In recent years, higher average temperatures during the receding winter in temperate areas is a major cause of concern among the fruit growers as it leads to early bloom and maturity.

9.1.3 Effect on Occurrence of Diseases and Pests

- Wetter, warmer winters favor incidence of diseases such as *Phytopthora*, *Fusarium*.
- Drier, warmer summers favor diseases such as *Alternaria*, *Powdery Mildew*, *Sooty bark diseases* under moisture stress.
- Warm and humid temperature favors Apple scab.
- Very dry summers result in increase in population of red mites and aphids, which makes more problem of attack on apple as shown in Fig. 9.2.

9.1.4 Effect on Pollination

9.1.4.1 Temperature

- If the temperature is either very low or very high, there is no fertilization, affecting fruit set, flower size, time of anthesis, and nectar pollen production in different crops.
- Most of the insects work well at or near 40 °F, and when the temperature is either low or very high, they do not take flight, which affects pollination and thereby fruit set rainfall and wind.
- Rainfall during flowering time affects the activity of pollen-carrying insects.
- Pollen-carrying insects work more effectively in a still atmosphere.

9.1.4.2 Relative Humidity

- Activity of bees and other pollen-carrying insects is hindered under low or very high relative humidity.

Fig. 9.2 Occurrence of diseases and pests

9.1.5 Effect on Dormancy and Chilling Requirement

- Dormancy is a mechanism that plants use to protect sensitive tissue from unfavorable conditions.
- The rapid climatic changes might alter the adaptability of many temperate fruit crops in the near future and severe production problems might arise.
- Commercially successful cultivars of many fruit crops require the fulfilment of winter chilling requirement, which is specific for every tree cultivar.
- Lack of chilling as in mild winter conditions results in abnormal pattern of bud-break and development in temperate fruit crops.
- In temperate climate areas, frost can also represent the main cause of weather-related damage to crops. Apples and other temperate fruit crops are vulnerable to spring frost.

9.1.6 Effect on Phenology

- The change in the timing of different physiological activities, i.e., phenology, is one of the impacts of climate change, as earlier date of full bloom of up to 10 days was observed in the apple varieties such as "Buskoop," "Cox's orange pippin," and "Golden Delicious" when comparing the last 20 years with the previous 30 years, which is less than the 40 days generally reported for Germany.

9.1.7 Mitigation Measures

- Climate change impacts are to be looked not in isolation but in conjunction with all the aspect of agriculture and allied sectors.
- Effects of climate change on horticulture sector are still uncertain.
- In the light of possible global warming, researchers should give more emphasis on development of heat and drought resistance crops.
- Development of "low chilling cultivars" in fruit plants is needed, some of which are mentioned in Table 9.1.
- Research is needed to define the current limits to this resistance and the feasibility of manipulation by modern genetic techniques.
- Both the crop architecture and physiology may be genetically altered to adopt to warmer environment conditions.
- At the regional level, those charged with planning for resource allocation, including land, water, and agriculture development, should also take climate change into account.
- Continuation of current and new initiatives of research is required so as to minimize the effects of climate change at farm, regional, national, and international level. It will help to provide a detailed picture of how world horticulture and agriculture could change.
- Formulation and implementation of policy and other adoptions in horticultural systems would minimize the negative effect.

Table 9.1 Low chill cultivars of some fruit crops

Fruit crops	Low chill cultivars	Chilling requirement
Apple	Anna, Mayan, Tamma, Tropical Beauty, Parlins Beauty, Neomi	<800 h
Peach and nectarines	Flordasun, Flodared, Sun Red nectarine, Sun Gold nectarine	<500 h
Pear	Gola, Pathernek (sand pear), Punjab Nectar	<150 h
Plum	Santa Rosa, Jamuni, Alucha Purple	–
Apricot	New Castle, Early Shipley, Keisha	–

9.2 Climate Change on Vegetable Cultivation

The key issues are:

- Mean annual temperature in India has increased over a period of time due to increase in CO_2 and CH_4 in atmosphere.
- Potato productivity is expected to decline in all potato-growing states of India and is highest for Karnataka, with a reduction of 18.68% by 2020, because potato crop is sensitive to temperature and an increase in temperature affects its tuberization.
- Fecundity of pests and migration of many bird species also increase due to rise in temperature and other related factors.

9.2.1 Climate Change vs Vegetables

Vegetables are generally more sensitive to environmental factors, and thus high temperatures and limited soil moisture are the major causes of low yields as they greatly affect several physiological and biochemical processes such as reduced photosynthetic activity, altered metabolism and enzymatic activity, thermal injury to the tissue, reduced pollination, and fruit set, which will be further magnified by climate change. In addition to physiological and chemical changes, many environmental factors such as temperature also increase insect, pest, and disease incidence. Development of plant diseases depends largely on environment prevailing around the host and pathogen, and a change in the constituents may influence host susceptibility and consequently host-parasite relationship.

9.3 Climate Change on Vegetable Crops

9.3.1 Temperature

Temperature fluctuations is primary the effect of climate change that adversely affects production of vegetables because physiological, biochemical, and metabolic activities are temperature dependent. Potato is well known for its exact temperature and day length requirement for tuber formation as well as flowering, so it becomes the most vulnerable crop for climate change. At low temperatures, starch is converted into the sugar, which causes browning due to charring of sugar while manufacturing chips, reducing their preference by the processing industry. This ultimately results in increased postharvest losses more than the present level, which is figured as 40–50%. The optimum temperature for development of lycopene pigment in tomato is 21–24 °C. The degradation of lycopene starts at above 25 °C and it is completely destroyed at 40 °C. The increase in temperature also affects pollination process.

9.3.2 Drought and Salinity

Drought and salinity and are the most important side effects of global warming. The prevalence of drought conditions adversely affects the germination of seeds in vegetable crops such as onion and okra and sprouting of tubers in potato. Potato is highly sensitive to drought and water stress, which results in poor tuberization and reduction in tuber yield. Drought increases the salt concentration in the soil and affects the reverse osmosis of loss of water from plant cells. This leads to an increased water loss in plant cells as well as the inhibition of several physiological and biochemical processes such as photosynthesis and respiration, reducing productivity of most vegetables.

Salt stress causes:

- Reduction in growth
- Wilting
- Leaf abscission
- Decrease photosynthesis and respiration
- Tissue necrosis and finally death.

Onions are susceptible to saline soils, while cucumber, eggplant, pepper, and tomato are moderately sensitive to saline soils.

9.3.3 CO_2 and Relative Humidity

Due to increased anthropogenic activities, the concentration of greenhouse gases such as CO_2 and CH_4 is increasing in the atmosphere day by day. They are not only responsible for global warming but also cause their own direct effect on growth and development of plants. Potato plants grown under elevated CO_2 may have larger photosynthetic rates up to some extent, but later on with increase in CO_2 concentration the photosynthetic rates decrease. Relative humidity and CO_2 can potentially affect pest and disease occurrence. High atmospheric CO_2 levels increase caterpillar food consumption, aphid reproduction, and decrease nitrogen-based plant defense; the beneficial effects are increased predation by predators. Precipitation washes away some insects such as whiteflies and thrips from fields, but high humidity due to high precipitation favors some insects such as tomato and legume pod borer (*Helicoverpa armigera*) and okra and eggplant leaf hopper (*Amrasca biguttula biguttula*). However, high humidity also favors the fungal pathogens of insects.

9.3.4 Air Pollutants and UV Radiations

Increased SO_2 and O_3 levels of 50–100 ppb increased the severity of fungal and nematode diseases by stimulating spore germination and invasion of fungi. Increase in gases such as sulfur dioxide and oxides of nitrogen in the atmosphere causes degradation of ozone layer, leading to penetration of harmful rays, i.e., UV rays. Vegetables such as tomato, cabbage, potatoes, and sugar beets are more susceptible to UV radiation than others.

9.4 Climate Change on Plantation Crops

India's coffee and tea growers are realizing that climate change—long spells of exceptionally warm weather, rain concentrated over short rather than long periods, and extreme weather events—can no longer be wished away. Rising overall output figures mask both regional variations and impact on quality. India registered a record tea output in 2015–16 despite a 7% decline in South India's gardens. The possibility of failure of both southwest monsoon and northeast monsoon can lead to further decline of tea quality. Tea quality has been a concern in Assam, which accounts for half of India's tea output. Harsh temperatures in the middle of the year and a tendency for the southwest monsoon to set in late have robbed Assam tea of some of its flavor. The absence of well-distributed rain, along with the rise in dry spells and temperatures, has led to more pest attacks. This has contributed to higher costs both to contain pests and to keep irrigation going in the drier months, squeezing margins in an increasingly competitive global market where India is losing ground to Sri Lanka and Kenya. To keep Indian teas alive both at home and overseas, it is important to adjust cropping practices. Micro-application of nutrients and rainwater harvesting can help beat back vagaries of the weather. However, both State and Central Governments need to play an active role in pushing this change by providing capital and know-how. Creating shade conditions for tea will help, but beyond a point can trigger fungal attacks.

Coffee, which too needs well-distributed rain, is under stress, even though Indian varieties are primarily shade-grown, unlike in Brazil and Vietnam. Shade coffees offer less yields than open varieties, but the trees last longer and are subject to less stress. They impart the flavor for which India's Arabica and Robusta coffees are so well regarded around the world. Even so, India's coffees have to cope with higher swings of temperature than in the past, with two consecutive drought years in Karnataka exposing the crop to white stem-borer disease. Coffee being a flowering plant, it critically depends on pollination in the blossom season in March, during which both bee activity and timely rains are critical. With climate change, a timely shower can no longer be relied on. Bee activity, too, has been affected by changing weather patterns, for which pesticide-chemical usage too is responsible. Both coffee and tea cultivation must shift toward more organic practices, and actively engage in the protection of biodiversity, not just to sustain itself but also to carve a niche in the world market. Climate change threatens to ramp up costs for agriculture as a whole, for which the government needs to adopt a well-thought-out response. This includes promoting research on drought and flood-resistant varieties, shorter duration crops, and crop diversification. Ecology and economics must go hand in hand in the plantation sector and elsewhere.

9.4.1 Relationship Between Yield and Climate

The impact of climate change on tea production of South Indian tea-growing regions was established in the section using observed values. Preliminary investigation on mean yield of all the study area showed that in the last two decades, the yield recorded negative trend (Fig. 8.4). Therefore, 20 years of climatic and productivity data (1991–2010) have been used in the present study.

9.5 Mitigation Strategies and Prospectus of Genetic Manipulation of Crops

Climatic variability and global warming observed during the past few decades have become a frightening reality and a matter of deep concern worldwide, especially for the developing countries. The extreme hot and cold seasons and subsequent drought and flood as well as the elevated greenhouse gases are generally considered to be the major causes of such global climatic variability. Continued industrialization aggravated by anthropogenic activity has increased the severity of climate change. Extreme climatic conditions adversely affect quantitative and qualitative characteristics of plants. Food and nutritional security have emerged as a major challenge worldwide due to overexpansion of population and cultivated areas to less fertile fields as a result of extreme climatic changes. Therefore, there is an urgent need to find out ways to mitigate the negative effects of climate change on biodiversity and crop productivity.

9.5.1 Mitigation

Mitigation involves attempts to slow the process of global climate change, usually by lowering the level of greenhouse gases in the atmosphere—another term for dealing with the causes rather than the consequences of the climate change problems. In other words, the goal is to avoid human interference with the climate system. The aim is to prevent the release of large quantities of CO_2 into the atmosphere from heavy industry.

The key measures according to IPCC (Intergovernmental Panel on Climate Change) for climate mitigation strategies are as follows:

(a) **Zero and low carbon energy sources**: Renewable energy (wind energy, solar energy, tidal energy, hydrothermal energy), nuclear energy (nuclear power), and bioenergy (biofuels).
(b) **Carbon capture and storage (CCS) (or carbon sequestration)**: It is the process of capturing waste carbon dioxide (CO_2), usually from large point sources, such as a cement factory or biomass power plant, transporting it to a storage site, and depositing it where it will not enter the atmosphere. Carbon dioxide can be captured from the industrial source (fossil fuel-based power

plants) or directly from the air. Carbon sequestration from industrial source involves the following.

(c) **Geo-sequestration**
Geological sequestration refers to the storage of CO_2 underground in depleted oil and gas reservoirs, saline or deep, unminable coal beds, or injected into porous rock formations. In depleted oil and gas reservoirs, it is used to produce more oil, called as enhanced oil recovery, because it reduces oil viscosity and improves oil displacement. Unminable coal seams can be used to store CO_2 because CO_2 molecules attach to the surface of coal.

(d) **Ocean sequestration**
The concept of storing carbon dioxide in the ocean was first proposed by the Italian physicist Cesare Marchetti in his 1976 paper "On geo-engineering and the carbon dioxide problem." Since then, the concept of sequestering atmospheric carbon dioxide in the world's oceans has been investigated by scientists, engineers, and environmental activists. Currently, 39,000 GtC (giga tons of carbon) reside in the oceans, while only 750 GtC are in the atmosphere. Carbon stored in oceans through direct injection. The ocean is a large natural sink for CO_2 and spontaneously absorbs a significant fraction of the fossil CO_2 emissions. After atmospheric carbon dioxide dissolves into the ocean, the aqueous carbon dioxide reacts with seawater to form carbonic acid. However, it is neutralized by calcium bicarbonate formed as discussed below. Following injection, waters interact and mix based on density and dilute the concentration of carbon dioxide, resulting in "carbon mineralization" and the reaction occurs as follows:

$$CaCO_3 + CO_2 + H_2O \rightarrow Ca^{2+} + 2HCO_3^-$$

(e) **Terrestrial sequestration**
Terrestrial sequestration (sometimes termed "biological sequestration") is typically accomplished through forest and soil conservation practices that enhance the storage of carbon (such as restoring and establishing new forests, wetlands, and grasslands) or reduce CO_2 emissions (such as reducing agricultural tillage and suppressing wildfire). The plants and the biomass act as a natural sink for carbon storage, increasing carbon fixation through photosynthesis.

9.5.2 Agriculture

Sustainable agriculture or regenerative agriculture in which organic substances are encouraged is the best way to mitigate the climate change. Modification of agricultural practices is a recognized method of carbon sequestration as soil can act as an effective carbon sink. Carbon emission reduction methods in agriculture can be grouped into two categories: reducing and/or displacing emissions and enhancing carbon removal.

9.5.2.1 Reducing Emissions

Increasing yields and efficiency generally reduces emissions as well, as more food results from the same or less effort. The techniques include more accurate use of fertilizers, less soil disturbance, better irrigation, and crop strains bred for locally beneficial traits and increased yields. In practice, most farming operations that incorporate postharvest crop residues, wastes, and by-products back into the soil provide a carbon storage benefit.

9.5.2.2 Enhancing Carbon Removal

All crops absorb CO_2 during growth and release it after harvest. The goal of agricultural carbon removal is to use the crop and its relation to the carbon cycle to permanently sequester carbon within the soil. This is done by selecting farming methods that return biomass to the soil and enhance the conditions in which the carbon within the plants will be reduced to its elemental nature and stored in a stable state. Methods for accomplishing this include the following.

Conservation Tillage

This method comprises low till/no-till farming—minimizing or eliminating manipulation of the soil for crop production. This includes the practice of mulch tillage, which leaves crop residues on the soil surface. These procedures generally reduce soil erosion, improve water-use efficiency, and increase carbon concentrations in the topsoil. Conservation tillage can also reduce the amount of fossil fuel consumed by farm operations.

Cover Cropping

This method comprises the use of crops such as clover and small grains for protection and soil improvement between periods of regular crop production. Cover crops improve carbon sequestration by enhancing soil structure and adding organic matter to the soil.

Crop Rotation

Planting different crops in a rotating pattern of years (e.g., corn-oats-clover) will reduce the loss of carbon from the soil and, with some inputs (e.g., manure-lime-phosphorous), will add carbon to soils.

Wetland Restoration

Wetland soil is an important carbon sink: 14.5% of the world's soil carbon is found in wetlands, while only 6% of the world's land is composed of wetlands.

9.6 Genetic Manipulation of Crops to Maximize Crop Production Under Climate Change Scenario

The works related to biotechnology strategies such as breeding and genetic engineering are the ways to maximize crop production under climate change scenario.

9.6.1 Breeding

It involves crossing plants to combine/enhance desirable traits/characters. It takes many generations to breed. Modern approaches use molecular gene markers to be able to trace interesting gene regions during breeding.

9.6.2 Breeding for Heat-Tolerant Crops

Tolerance to heat stress is an important and desired trait and several seed companies make efforts to develop thermotolerant cultivars. This is essential not only for crops cultivated exclusively in the field such as rice, wheat, and maize but also for field-grown vegetables such as tomato. In rice, a recombinant inbred line (RIL) population was screened for thermotolerance based on spikelet fertility at high temperatures. They identified eight loci mapped to different chromosomes that were associated with spikelet fertility, while Quantitative Trait Locus (QTLs) from similar chromosomal positions have also been associated with tolerance to other stresses including drought, cold, and salinity, indicating that these loci are probably involved in common responsive genomic regions. Similarly, QTLs related to heat and drought tolerance have been identified in wheat. In maize, it was found that QTLs related with heat stress tolerance by screening pollen germination rate and pollen tube growth, while scientists identified two QTLs that controlled grain filling under stress conditions. Genotyping for heat stress tolerance has led to the selection of populations that are suitable for cultivation and fruit production in environments with higher temperatures. The selection is based on several phenotyping factors, including days to 50% flowering, flowers per cluster, percentage of fruit set, number of fruit clusters per plant, fruits per plant, individual fruit weight, fruit yield per plant, fruit length and diameter, seeds per fruit, and percentage viable pollen grain, all of which were measured in tomato. Therefore, a successful breeding program highly depends not only on the genetic material used but also on the screening methods applied due to the polygenic nature of heat stress response and thermotolerance. Various varieties of different crops have been developed through breeding programs that are resistant to abiotic stresses such as droughts and salinity and biotic stresses such as diseases and insect pests.

9.7 Genetic Engineering

Genetic engineering is the process in which a desired gene is isolated, cloned, and inserted into a host organism. Genetic engineering allows the introduction of genes from genetically distant sources, overcoming the problem of cross-barrier species, distant relatives, or even transformation of genes from non-plant organisms.

9.7.1 Enhancing Heat Tolerance by Genetic Engineering

Introduction of SAMDC from *S. cerevisiae* to tomato enhanced polyamine biosynthesis, antioxidant enzyme activity, and CO_2 assimilation, resulting in increased thermotolerance of the transgenic plants. Similarly, expression of maize EFTu1 in wheat reduced protein aggregation in heat-stressed leaves and injury to photosynthetic membranes, enhanced the rate of CO_2 fixation, and improved the tolerance to heat stress. In another approach, several transgenic potato lines expressing Hsp17.7 from carrot were used to make the cold crop plant more heat resistant. The different potato lines showed increased thermotolerance, measured by higher membrane stability as well as improved tuberization at high temperatures.

9.8 Modification of RuBisCO as a Result of Climate Change

Global warming and the rise in atmospheric CO_2 will increase the operational temperature of leaves in next decades, typically well above the thermal optimum for photosynthesis. Presently, there is controversy over the limiting processes that control photosynthesis at elevated temperature. Leading models propose that the reduction in photosynthesis at elevated temperatures is a function of either a decline in the capacity of electron transport to regenerate RuBP or a decrease in the capacity of RuBisCO activase to maintain RuBisCO in an active configuration.

Bibliography

Cheng H, Zhang PZ, Spötl C, Edwards RL, Cai YJ, Zhang DZ, Sang WC, Tan M, An ZS (2012) The climatic cyclicity in semiarid-arid Central Asia over the past 500,000 years. Geophys Res Lett 39(1):L01705

Chung BN, Canto T, Tenllado F, San Choi K, Joa JH, Ahn JJ, Kim CH, Do KS (2016) The effects of high temperature on infection by potato virus Y, potato virus A, and potato leafroll virus. Plant Pathol J 32(4):321

Frova C, Sari-Gorla M (1994) Quantitative trait loci (QTLs) for pollen thermotolerance detected in maize. Mol Gen Genet MGG 245(4):424–430

Hanif U, Syed SH, Ahmad R, Malik KA, Nasir M (2010) Economic impact of climate change on the agricultural sector of Punjab [with comments]. Pak Dev Rev 49:771–798

Hansen JW (2002) Realizing the potential benefits of climate prediction to agriculture: issues, approaches, challenges. Agric Syst 74(3):309–330

Houghton JT, Jenkins GJ, Ephraums JJ (1990) Climate change: the IPCC scientific assessment, vol 80. Cambridge University Press, Cambridge

Nagarajan S, Jagadish SVK, Prasad AH, Thomar AK, Anand A, Pal M, Agarwal PK (2010) Local climate affects growth, yield and grain quality of aromatic and non-aromatic rice in northwestern India. Agric Ecosyst Environ 138(3–4):274–281

Sage RF, Way DA, Kubien DS (2008) Rubisco, rubisco activase, and global climate change. J Exp Bot 59(7):1581–1595

Seidel DJ, Fu Q, Randel WJ, Reichler TJ (2008) Widening of the tropical belt in a changing climate. Nat Geosci 1(1):21–24

Soto-Pinto L, Anzueto M, Mendoza J, Ferrer GJ, de Jong B (2010) Carbon sequestration through agroforestry in indigenous communities of Chiapas, Mexico. Agroforestry Systems 78(1): 39–51

Climate Change Effect on Forests 10

Abstract

Climate impacts the function of forest ecosystems and plays a major role in forest health. Climate and forests are inherently linked. Climate change can affect the forest productivity both directly and indirectly. These changes are having significant influence on the world's forests and the forest sector, which acts as both a cause and an effect of our changing climate, e.g., through longer growing seasons, shifts in insect species ranges, forest loss, and degradation. Besides direct effects, climate change indirectly can also change the composition and dynamics of the entire forest ecosystem. One of the promising solutions to address the effect of climate change is forests. This chapter introduces all the facts related to climate impact on forests and their productivity. In this chapter, various other related topics has also been discussed, which mainly include UN convention on climate change—NATCOM initiatives, Kyoto Protocol, National Action Plan on Climate Change, Green India Mission, Indian Network for Climate Change Assessment (INCCA), State Action Plans on Climate Change.

Keywords

Global warming · Vulnerability · Adaptability · Agroforestry · NATCOM · Kyoto Protocol · INCCA · SAPCC

Living and nonliving organisms, their ecosystems, the environment, and climatic conditions can be considered as dependent components of the world on each other and their balance is essential for the sustainability and stability of the world. According to the Intergovernmental Panel on Climate Change (IPCC) report, global average temperature of the period 1880–2012 has increased by 0.85 °C and the temperature increase between the 1850–1900 period and the 2003–2012 period was 0.78 °C. Consequently, the rate of sea level rise was also high (annual rate of

L. Ahmad et al., *Climate Change and Agrometeorology*,
https://doi.org/10.1007/978-981-99-4863-5_10

3.2 mm/year). Furthermore, carbon dioxide (CO_2) concentration has increased by more than 25% during the last century, and according to the IPCC fifth assessment report of climate change, CO_2 concentration has increased by 40% primarily due to two anthropogenic processes: fossil fuel emission and deforestation (land use change).

10.1 Global Warming: Its Causes and its Effects

The growing accumulation of greenhouse gases in atmosphere contributes much of the global warming as a result of anthropogenic activities, primarily economic activities. Carbon dioxide is the major and primary greenhouse gas accumulating in the atmosphere and its main cause is the continuing global deforestation and other land use changes that add to these emissions. The examples of land use changes include loss of other vegetation and depletion of organic matter in soil.

There is no doubt that these recent climate changes can intrinsically affect forest ecosystems. Severe climate change is one of the causes of deforestation, which can lead to desertification, while at the same time this deforestation is a major driver of climate change. The increasing temperature and drought have negative impacts on species diversity as well as ecosystem goods and services to humanity.

10.2 Deforestation and Its Effects on Climate Change

The major drivers of climate change can be divided into two types: anthropogenic causes such as land use changes, deforestation and forest degradation, fossil fuel burning, and industrial processes; and natural causes such as solar radiation changes and volcanic activity. So, deforestation and forest degradation play a role in global climate change, and reciprocally, deforestation is indirectly affected by climate change.

About 10,000 years ago, the world's forests covered around 6 billion hectares, which is about 45% of total land area. However, the forest cover has been decreasing from that period and only about 31% of the world's area is covered by forests in 2010. Agricultural expansion including shifting cultivation is one of the major anthropogenic divers for deforestation in the tropical region. According to the reports of the Forest and Agriculture Organization (FAO), annual rate of deforestation is 0.14% per year (5.2 million hectares per year of net annual deforestation) and, during the twenty-first century, deforestation was slow in temperate forests, whereas deforestation was still high in tropical forests. Globally, greenhouse gas (GHG) emissions from deforestation and forest degradation have been found to account for about one-fifth of the total GHG emissions. Evidence of vast impacts on climate change by CO_2 emission from deforestation was found in Indonesia and Brazil, which are the third and fourth greatest CO_2 emitters of the world. Hence, deforestation and forest degradation are vital considerations for sound environmental management because they significantly effect on climate change and contribute to the global warming process.

10.3 Impacts of Climate Change on Forests

10.3.1 Forest Growth

Tree and forest growth rate is significantly correlated with climate. The growth rate is strongly influenced by water availability. If there is sufficient water availability, a short and not very intense dry period, and high temperature, a high growth rate will be found. Solar radiation is another factor that affects tree growth rate. However, the effect on tree growth rate is more significant if solar radiation effect is combined with temperature effect. The variation in solar radiation and temperature together affected tree growth rate at stand level although each factor separately did not affect tree growth rate. Furthermore, the collective influence of solar radiation, precipitation, soil water content, and temperature upsurge contributes to the increased tree growth variation.

10.3.2 Forest Regeneration

Climate changes such as high range of temperature shift and high variation in water availability can limit the regeneration capacity of tree species because tree seedlings are very susceptible to climate change. In temperate forests, seedling mortality rate of *Aextoxicon punctatum* was significantly affected by reduced precipitation and increased temperature.

Drought can be considered as one of the limitations for seedling establishment. Drought-induced canopy losses of *Pinus sylvestris* can reduce understory *Quercus* spp. germination although they can favor the growth of light-demanding species that have already established before drought. Furthermore, one of the consequences of climate change, fire, can affects seedling density. In tropical moist deciduous forest, the seedling density of burned forest was 63% lower than unburned forest.

10.3.3 Species Composition and Diversity

Drought and heat impacts of climate change can cause tree mortality that alters species composition and diversity. Likewise, species distribution and migration are directly affected by fire responses to climate change. In tropical lowland and moist forests, water deficiency increased the rate of tree mortality and this effect was severe by increasing frequency and rate of water deficiency. However, tropical moist forests have a greater tendency to resist water shortage compared with tropical dry forests. Therefore, species mortality through water deficiency was found to occur more in the tropical dry forests, and it severely affected trees of small diameter.

Another factor that can change the species composition and stand structure is fire. In both tropical moist and dry deciduous forests, seedling and sapling densities decreased and species diversities decreased by 50% and 60% by moderate and intense fire frequency, respectively, when compared with low fire frequency.

10.3.4 Forest Habitat and Forest Ecosystems

A habitat can be defined as a natural environment or natural region that is inhabited by particular species or living organisms and an ecosystem consists of abiotic and biotic components that depend on each other and exist together in balance with the corresponding surrounding environment. Forest habitats and forest ecosystems respond to climate change. According to one scientist who used an inventory and model, montane conifer (*Pinus sylvestris*), sub-Mediterranean conifer (*Pinus nigra*), and sub-Mediterranean deciduous (*Quercus pyrenaica*) oak are facing high vulnerability to climate change, whereas low vulnerabilities to climate change was found in *Quercus robur*, *Quercus petraea*, *Pinus pinaster*, *Quercus faginea*, and *Quercus ilex* species. However, some species, *Pinus pinea* and *Pinus halepensis*, have been found an increase in their frequency of occurrence with climate change. By simulation analysis, the number of trees of *Pinus halepensis* species will increase 50% with climate change in comparison with a stable climate by 2100.

10.4 Forest and Climate Change

The impacts of climate change are already evident on the vegetation systems in the Himalayas. During the last century, global average surface temperature has increased by approximately 0.6–0.8 °C. Nonetheless, during past decade the rates of climatic warming in Himalaya have been reported to exceed over 1.2 °C, which has drastically impacted the forest ecosystems in this region. Various studies have reported the direct and indirect role of climate change in vegetation degradation such as invasion of alien species, forest destruction due to increased occurrences of forest fires, and reductions in agricultural and forest productivity.

10.4.1 Climate Change Vulnerability

Vulnerability is the degree to which a system is susceptible to, or unable to cope with, impacts of climate change, including climate variability and extremes. IPCC defines it as an integrated measure of the expected magnitude of adverse effects to a system caused by a given level of certain external stressors. The term "vulnerability" has been used by various research communities in many different ways, such as those focused on securing livelihoods, food security, natural hazards, disaster risk management, public health, global environmental change, and climate change. Scientists reported that "climate change vulnerability of forests can be related to concepts such as resilience, marginality, susceptibility, adaptability, fragility, and risk." However, climate change vulnerability of forests is also a function of exposure, sensitivity, and adaptive capacity of these ecosystems in a particular region, i.e., Vulnerability = f(Exposure, Sensitivity, Adaptive capacity).

As per IPCC's definition, vulnerability includes an external dimension, which is represented here by the "exposure" of a system to climate variations, as well as an internal dimension, which comprises its "sensitivity" and "adaptive capacity" to these stressors. Exposure can be defined as the "nature or the degree to which a system is exposed to significant climatic variations," and it comprises climatic variables derived from temperature and precipitation. Sensitivity is defined as the degree to which a system is affected either adversely or beneficially by climate-related stimuli, and it contains biophysical variables that indicate the phonological and physiological characteristics of the forests, e.g., albedo, evapotranspiration, occurrence of forest fires, net primary productivity, and gross primary productivity. Adaptive capacity can be defined as ability of a system to adjust to climate change to moderate potential damage from it, and it comprises variables pertaining to plant species richness and forest structure.

10.4.2 Ecosystem Response to Climate

Climate change affects a wide range of ecosystem functions and processes. Projecting the fate of forest ecosystems under a changing climate is more challenging than for species. These include direct effects of temperature and precipitation on physiological and reproductive processes such as photosynthesis, water use, flowering, fruiting and regeneration, growth and mortality, and litter decomposition. Changes in these processes will have effects on species attributes such as wood density or foliar nutrient status. Indirect effects will be exhibited through changing fire and other climate-driven disturbances. These will ultimately have impacts on stand composition, habitat structure, timber supply capacity, soil erosion, and water yield.

Forest ecosystem responses to elevated CO_2 and climate change is given as follows:

Direct	Faster response	Slower response
	Photosynthesis	Litter decomposition
	Water use/transport	Foliar nutrient status
	Flowering/phenology	Genetic change
	Regeneration	Tree species distribution and stand composition
	Growth and mortality	
	Wood density/quality	
	Frost/storm damage	
Indirect	Increased fire frequency/intensity	Habitat composition and structure
	Disease impacts	Wood supply
	More insect pests	Erosion and soil loss
	Increased invasive species	Water yield

(continued)

Direct	Faster response	Slower response
	Photosynthesis	Litter decomposition
	Water use/transport	Foliar nutrient status
	Flowering/phenology	Genetic change
	Regeneration	Tree species distribution and stand composition
	Growth and mortality	
	Wood density/quality	
	Frost/storm damage	
Indirect	Increased fire frequency/intensity	Habitat composition and structure
	Disease impacts	Wood supply
	More insect pests	Erosion and soil loss
	Increased invasive species	Water yield
	Water quality	

10.4.3 Fire, Pests, Invasive Species, and Disturbance Risks

10.4.3.1 Fire

The frequency, size, intensity, seasonality, and type of fires depend on weather and climate in addition to forest structure and composition. Fire initiation and spread depend on the amount and frequency of precipitation, the presence of ignition agents, and conditions (e.g., lightning, fuel availability and distribution, topography, temperature, relative humidity, and wind velocity).

Fire effects on forests include acceleration of nutrient cycling, mortality of individual trees, shifts in successional direction, induced seed germination, loss of soil seed bank, increased landscape heterogeneity, changes in surface-soil organic layers and underground plant root and reproductive tissues, and volatilization of soil nutrients. Erosion can occur where soil disturbance accompanies fire (e.g., during firefighting or timber salvage operations). Fire affects forest value for wildlife habitat, timber, recreation, and through smoke and human health.

10.4.3.2 Drought

Droughts occur in nearly all forest ecosystems. Drought effects are influenced by soil texture and depth, exposure, species present, life stage, and the frequency, duration, and severity of drought. Droughts occur irregularly in forests of the humid regions east of the Mississippi River and in the super humid Pacific Northwest. Droughts occur annually at the end of the growing season in forests at the mid-continental prairie–forest border, where annual precipitation ranges from 600 to 1000 mm or within humid regions that have shallow or rocky soils.

10.4.3.3 Insect and Pathogen Outbreak

Climate influences the survival and spread of insects and pathogens directly, as well as the susceptibility of their forest ecosystems. Changes in temperature and precipitation affect herbivore and pathogen survival, reproduction, dispersal, and distribution. Indirect consequences of disturbance from herbivores and pathogens include elimination of nesting trees for birds and negative effects on mycorrhizal fungi. Other indirect effects include the impacts of climate on competitors and natural enemies that regulate the abundance of potential pests and pathogens.

Changes in the intensity and frequency of herbivore and pathogen damage in forests can have a range of effects. Most tree species support a community of other organisms, so the loss of any tree species can significantly reduce overall biodiversity. Such a loss occurred when chestnut blight almost completely eliminated chestnut trees; the die-off of Fraser fir (*Abies fraseri*) caused by balsam woolly adelgid (*Adelges piceae*) also raises concerns. Herbivore and pathogen damage to trees can increase understory plant diversity; the overall abundance and diversity of birds; and the diversity of predators, parasitoids, and detritivores.

In addition, as insects typically migrate much faster than trees, many temperate tree species are likely to encounter non-native insect herbivores that previously were restricted to subtropical forests.

10.4.3.4 Introduced Species

Introduced species can affect forests through herbivory, predation, habitat change, competition, alteration of gene pools via hybridization with natives, and disease (as either pathogens or vectors). Introduced species can alter the diversity, nutrient cycles, forest succession, and fire frequency and intensity of some ecosystems. The effects of introduced species should be considered concurrently with changes in native species distribution and abundance that occur as a consequence of climate change. The impact of introduced species on ecosystems is influenced by such climatic factors as temperature, drought, and cloud cover.

The ultimate ranges of introduced species are largely determined by climate and human activities. Climate change will modify the distributions of many introduced species. Developmental rates will be modified by temperature change. For example, laboratory studies of balsam woolly adelgid growing under various temperature conditions provided the basis for simulations that suggest that temperature-induced changes in the population dynamics of the insect significantly affect Fraser fir survival. Increased CO_2 can directly influence introduced plants through enhanced photosynthesis, but at different rates for different species. Resistance of trees to introduced herbivores is sensitive to both climate and CO_2 concentrations. Climate change, in concert with CO_2 concentration and nitrogen deposition, affects leaf nitrogen, which in turn influences herbivory.

10.4.4 Mitigation and Adaptation

Two fundamental response options to the predicted climate change are mitigation and adaptation to the climate change, respectively. Mitigation refers to controlling global climate change by reducing the emission of greenhouse gases (GHGs) and enhancing their sinks, whereas adaptation mainly focuses on moderating impacts of climate change through a wide range of actions that are targeted at the vulnerable system.

Climate change presents significant potential risks to forests and challenges for forest managers. Adaptation to climate change involves monitoring and anticipating change and undertaking actions to avoid the negative consequences and to take advantage of potential benefits of those changes. While forest ecosystems are resilient and many species and ecosystems have adapted historically to changing conditions, future changes are potentially of such magnitudes or will occur at rates that are beyond the natural adaptive capacity of forest species or ecosystems, leading to local extinctions and the loss of important functions and services, including reduced forest carbon stocks and sequestration capacity.

Recent global warming has already caused many changes in forests. Aspects of climate change may be positive for some tree species in some locations. Tree growth is observed to be increasing in some locations under longer growing seasons, warmer temperatures, and increased levels of CO_2. However, many projected future changes in climate and their indirect effects are likely to have negative consequences for forests. Observed shifts in vegetation distribution or increased tree mortality due to drought and heat in forests worldwide may not be due to human-induced climate change but demonstrate the potential impacts of rapid climate change. These impacts may be aggravated by other human-induced environmental changes such as increases in low elevation ozone concentrations, nitrogenous pollutant deposition, the introduction of exotic insect pests and pathogens, habitat fragmentation, and increased disturbances such as fire. Other effects of climate change may also be important for forests. Sea level rise is already having an impact on tidal freshwater forests, and tidal saltwater forests (mangroves) are spreading landward in subtropical coastal reaches, taking over freshwater marsh and forest zones.

Adapting to these changing and uncertain future conditions can be considered from a number of perspectives. Policy and management might be directed at avoiding or reducing the impact of climate-related events, reducing vulnerability to future climatic conditions, managing a broader suite of climate "risks," or increasing resilience and capacity in forest ecological and production systems to recover from climate "shocks." Adapting forest management to climate change involves monitoring and anticipating change and undertaking actions to avoid the negative consequences or take advantage of potential benefits of those changes. Adopting the principles and practices of sustainable forest management (SFM) can provide a sound basis for addressing the challenges of climate change.

Ecosystem-based adaptation is being promoted as having the potential to incorporate sustainable management, conservation, and restoration of ecosystems into adaptation to climate change. This can be achieved more effectively by integrating ecosystem management and adaptation into national development policies through education and outreach to raise societal awareness about the value of ecosystem services.

Broadly, there has been little integration to date of mitigation and adaptation objectives in climate policy. For example, there is little connection between policies supporting the reducing emissions from deforestation and forest degradation plus

(REDD+) initiatives and adaptation. Integrating adaptation into REDD+ can advance climate change mitigation goals and objectives for sustainable forest management.

10.5 Climate Change Influence on Agroforestry

Climate change, as one of the biggest threats to a global food security, highly influences natural resources that are essential for crop production. Farming is not only affected by the impact of climate change, but it is also a significant contributor to global greenhouse gas emissions. Intensive farm practices that include both farm production and change of land use directly affect the global carbon, water, and nutrient cycles. Farming, as a direct driver for 80% of global deforestation, also releases significant amounts of greenhouse gasses. Furthermore, as forests continue to be cleared, the land is directly exposed to wind and water soil erosion. Agroforestry, as the bridge between forestry and farming, is a diversified set of farm production systems that integrates trees in the farms. There are three different types of agroforestry:

- Agri-silvicultural systems: the combination of crops and trees;
- Silvo-pastoral systems: the combination of forestry and grazing of domesticated animals; and
- Agro-silvo-pastoral systems: the combination of trees, animals, and crops.

Tree cover on farms has the potential to make an important contribution to climate change mitigation. Figure 10.1 shows the diagrammatic representation of sustainability of integrating trees in farming. There are many benefits of incorporating the planting of trees in farm management such as:

- Improve carbon sequestration: during the photosynthesis process, the tree absorbs carbon from the atmosphere. Studies have shown that a tropical tree absorbs about 22 kg of carbon each year. By using this practice, farmers can significantly mitigate climate change;
- Reduce soil erosion: farmers who plant trees as a natural barrier could prevent wind or water erosion;
- Improve soil management: deep-rooting trees improve soil stability; trees increase the soil's ability to absorb and retain water, produce nutrients, improve and maintain organic soil matter, and manage soil temperature;
- Trees attract vital pollinators and birds, which supports integrated pest management;
- Improve water management: surface water run-off is reduced, while trees protect nearby water courses by trapping farm pollutants; trees capture, filter, and store water and manage water supplies;
- Provide shelter and shade for livestock and crops;
- Lesser the risk of salinization;

Fig. 10.1 Sustainability of integrating trees in farming

- Provide additional revenue or food for farmers who grow trees that produce fruit, nuts, or leaves;
- The felled trees or their residues can be used as firewood;
- Leaves can serve as a fodder for livestock; and
- Certain trees are an important source of natural remedies.

10.5.1 Climate Resilient Forestry

Resilience is the capacity of a forest to withstand (absorb) external pressures and return, over time, to its pre-disturbance state. When viewed over an appropriate time span, a resilient forest ecosystem is able to maintain its "identity" in terms of taxonomic composition, structure, ecological functions, and process rates. Maintaining and restoring biodiversity in forests promotes their resilience to human-induced pressures and is therefore an essential "'insurance policy" and safeguard against expected climate change impacts. Biodiversity should be considered at all scales (stand, landscape, ecosystem, bioregional) and in terms of all elements (genes, species, communities). Increasing the biodiversity in planted and seminatural forests will have a positive effect on their resilience capacity and often on their productivity (including carbon storage).

The resilience of a forest ecosystem to changing environmental conditions is determined by its biological and ecological resources, in particular (1) the diversity of species, including microorganisms, (2) the genetic variability within species (i.e., the diversity of genetic traits within populations of species), and (3) the regional pool of species and ecosystems. The resilience is also influenced by the size of forest ecosystems (generally, the larger and less fragmented, the better) and by the condition and character of the surrounding landscape.

Primary forests are generally more resilient (stable, resistant, and adaptive) than modified natural forests or plantations. Therefore, policies and measures that promote their protection yield both biodiversity conservation and climate change mitigation benefits, in addition to a full array of ecosystem services. Nevertheless, it must be recognized that certain degraded forests, especially those with invasive alien species, may be stable and resilient, and these forests can become serious management challenges if attempts are made to reestablish the natural ecosystem to recover original goods and services. Some forest ecosystems with naturally low species diversity nevertheless have a high degree of resilience such as boreal pine forests. These forests, however, are highly adapted to severe disturbances, and their dominant tree species have a broad genetic variability that allows tolerance to a wide range of environmental conditions. The carbon pool is largest in old primary forests, especially in the wet tropics, which are stable forest systems with high resilience.

The permanence of efforts under UNFCCC (United Nations Framework Convention on Climate Change) negotiations, such as reducing emissions from deforestation and forest degradation (REDD), as well as other forest-based climate change mitigation and adaptation policies and measures, is linked to the resilience of forests, and hence forest biodiversity. Therefore, REDD activities should take biodiversity conservation into consideration, as this will help maintain forest ecosystem resilience and the long-term stability of the carbon pool.

The regional impacts of climate change, especially interacting with other land use pressures, might be sufficient to overcome the resilience of even some large areas of primary forests, pushing them into a permanently changed state. If forest ecosystems are pushed past an ecological "tipping point," they could be transformed into a different forest type and, in extreme cases, a new non-forest ecosystem state (e.g., from forest to savannah). In most cases, the new ecosystem state would be poorer in terms of both biological diversity and delivering ecosystem goods and services.

Plantations and modified natural forests will face greater disturbances and risks for large-scale losses due to climate change than primary forests because of their generally reduced biodiversity. The risks can partly be mitigated by adhering to a number of forest management recommendations:

- Maintain genetic diversity in forests by avoiding practices that select only certain trees for harvesting based on site, growth rate, or form.
- Maintain stand and landscape structural complexity, using natural forests and processes as models.

- Maintain connectivity across forest landscapes by reducing fragmentation, recovering lost habitats (forest types), expanding protected area networks, and establishing ecological corridors.
- Maintain functional diversity and eliminate the conversion of diverse natural forests to monotypic or reduced-species plantations.
- Reduce nonnatural competition by controlling invasive species and reduce reliance on non-native tree crop species for plantation, afforestation, or reforestation projects.
- Manage plantation and seminatural forests in an ecologically sustainable way that recognizes and plans for predicted future climates. For example, reduce the odds of long-term failure by apportioning some areas of assisted regeneration for trees from regional provenances and from climates that approximate future climate conditions based on climate modeling.
- Maintain biodiversity at all scales (stand, landscape, bioregional) and of all elements (genes, species, communities) by, for example, protecting tree populations that are isolated, disjunct, or at margins of their distributions, source habitats, and refuge networks. These populations are most likely to represent preadapted gene pools for responding to climate change and could form core populations as conditions change.
- Ensure that there are national and regional networks of scientifically designed, comprehensive, adequate, and representative protected areas. Build these networks into national and regional planning for large-scale landscape connectivity.

10.5.2 Economic Worth of Carbon Storage in Forest

One of the important ecological roles played by forests and one that currently provides the greatest potential to realize an economic return is carbon sequestration. The economic worth of carbon stored in the aboveground biomass of the forests is evaluated by equating it to the cost that would be incurred in offsetting by alternative projects, the carbon dioxide released if the forestlands were converted into alternate land use. A monetary value required to offset the additional carbon stores in the forests as compared to other land uses such as pasture of cropland is calculated. This monetary value is then converted into annual flows. The average cost (dollars/tons CO_2) of various other mitigation options is used to calculate the cost required to offset the CO_2 released by forestland conversion. The monetary value obtained is converted into annual flows by discounting it at market interest rate of 10%. This gives the worth of additional aboveground carbon stored in the forestland on annual basis.

10.6 UN Convention on Climate Change

In 1988, the IPCC was created by the World Meteorological Organization and the United Nations Environment Programme (UNEP) to study the evidence that concentrations of carbon dioxide in the atmosphere were increasing. This group issued a first assessment report in 1990, which reflected the views of 400 scientists. The report stated that global warming was real and urged that something should be done about it. The Panel's findings spurred governments to create the *United Nations Framework Convention on Climate Change* (UNFCCC) in 1992 to begin considering what could be done to reduce global warming and to cope with whatever temperature increases are inevitable. More recently, a number of nations approved an amendment to the treaty: the *Kyoto Protocol*, which has more powerful (and legally binding) measures. This was adopted in 1997 and is aimed at establishing a first step toward achieving the main objective of the Convention: to stabilize GHG emissions from human activities. The Protocol establishes emission targets for the Annex I Parties.

The Kyoto Protocol to the United Nations Framework Convention on Climate Change was adopted in Kyoto, Japan, in December 1997 and entered into force on February 16, 2005. Currently, there are 192 Parties to the Kyoto Protocol. The rules and requirements for implementation of the Kyoto Protocol were further elaborated in a package of decisions called the Marrakesh Accords. The Marrakesh Accords were formally adopted by the CMP (Conference of parties) at its first session in Montreal, Canada, in December 2005. The Kyoto Protocol shares the ultimate objective of the Convention to stabilize atmospheric concentrations of GHGs at a level that will prevent dangerous interference with the climate system. In pursuit of this objective, the Kyoto Protocol builds upon and enhances many of the commitments already in place under the Convention. Each Annex I Party has a binding commitment to limit or reduce GHG emissions, and innovative mechanisms have been established for Parties to facilitate compliance with this commitment. Other commitments include the following:

- Each Annex I Party must undertake domestic policies and measures to reduce GHG emissions and to enhance removals by sinks;
- In implementing these policies and measures, each Annex I Party must strive to minimize any adverse impact of these policies and measures on other Parties, particularly developing country Parties;
- Annex I Parties must provide additional financial resources to advance the implementation of commitments by developing countries;
- Both Annex I and non-Annex I Parties must cooperate in the areas of:
  (a) The development, application, and diffusion of climate-friendly technologies;
  (b) Research on and systematic observation of the climate system;
  (c) Education, training, and public awareness of climate change; and
  (d) The improvement of methodologies and data for GHG inventories.

In short, the Kyoto Protocol operationalizes the United Nations Framework Convention on Climate Change by committing industrialized countries and economies in transition to limit and reduce greenhouse gas (GHG) emissions in accordance with agreed-upon individual targets. The Convention itself only asks those countries to adopt policies and measures on mitigation and to report periodically. The Kyoto Protocol is based on the principles and provisions of the Convention and follows its annex-based structure. It only binds developed countries and places a heavier burden on them under the principle of "common but differentiated responsibility and respective capabilities," because it recognizes that they are largely responsible for the current high levels of GHG emissions in the atmosphere. In its Annex B, the Kyoto Protocol sets binding emission reduction targets for 37 industrialized countries and economies in transition and the European Union. Overall, these targets add up to an average 5% emission reduction compared to 1990 levels over the 5-year period 2008–2012 (the first commitment period).

10.6.1 NATCOM Initiatives

There are some observed changes in climate parameters in India. India's Initial National Communication, 2004 (NATCOM I) to UNFCCC has consolidated some of these. Some highlights from NATCOM I and others are listed as follows.

- **Surface temperature**
  At the national level, an increase in surface air temperatures of 0.4 °C has been observed over the last century. A warming trend has been observed along the west coast, in central India, the interior peninsula, and north-eastern India. However, cooling trends have been observed in northwestern India and parts of South India.
- **Rainfall**
  While the observed monsoon rainfall at the all-India level does not show any significant trend, regional monsoon variations have been recorded. A trend of increasing monsoon seasonal rainfall has been found along the west coast, northern Andhra Pradesh, and north-western India (+10% to +12% of the normal over the last 100 years), while a trend of decreasing monsoon seasonal rainfall has been observed over eastern Madhya Pradesh, north-eastern India, and some parts of Gujarat and Kerala (−6% to −8% of the normal over the last 100 years).
- **Extreme weather events**
  Instrument records over the past 130 years do not indicate any marked long-term trend in the frequencies of large-scale droughts and floods. However, trends are observed in multi-decadal periods of more frequent droughts, followed by less severe droughts. There has been an overall increasing trend in severe storm incidence along the coast at the rate of 0.011 events per year. While the states of West Bengal and Gujarat have reported increasing trends, a decline has been observed in Odisha. An analysis of a daily rainfall data set revealed (1) a rising trend in the frequency of heavy rain events and (2) a significant decrease in the frequency of moderate events over central India from 1951 to 2000.

- **Rise in sea level**
  Using the records of coastal tide gauges in the north Indian Ocean for more than 40 years, it has been estimated that sea level rise was between 1.06 and 1.75 mm/year. These rates are consistent with 1–2 mm/year global sea level rise estimates of IPCC.
- **Impacts on Himalayan glaciers**
  The Himalayas possess one of the largest resources of snow and ice and its glaciers form a source of water for the perennial rivers such as the Indus, the Ganga, and the Brahmaputra. Glacial melt may impact their long-term lean-season flows, with adverse impacts on the economy in terms of water availability and hydropower generation. The available monitoring data on Himalayan glaciers indicates that while recession of some glaciers has occurred in some Himalayan regions in recent years, the trend is not consistent across the entire mountain chain. Accordingly, it is too early to establish long-term trends, or their causation, for which there are several hypotheses exist.

10.7 Projections of Climate Change on Forests

Based on future climate, projections of Regional Climate Model of the Hadley Centre show that 77% and 68% of the forest areas in the country are likely to experience shift in forest types, by the end of the century, with consequent changes in forests produce and, in turn, livelihoods based on those products. As a result, the associated biodiversity is likely to be adversely impacted. The NATCOM I in India projects an increase in the area under xeric scrublands and xeric woodlands in central India at the cost of dry savannah in these regions.

10.7.1 Decision-Making in Emission of Greenhouse Gases (GHG)

Risk assessment and management provide a foundation for decision-making in considering climate change in natural resource management. This approach provides both a qualitative and quantitative framework for evaluating climate change effects and adaptation options. Incorporating risk management approaches into forest management plans can provide a basis for managers to continue to provide forest conditions that meet a range of important values.

10.7.2 National Action Plan on Climate Change

The National Action Plan on Climate Change (NAPCC), introduced in 2009, identifies measures that promote India's development objectives while also yielding co-benefits for addressing climate change effectively. It outlines a number of steps to simultaneously advance India's development and climate-change-related objectives of adaptation and mitigation.

The NAPCC is guided by the following principles:

- Protecting the poor and vulnerable sections of the society through an inclusive and sustainable development strategy, sensitive to climate change.
- Achieving national growth objectives through a qualitative change that enhances ecological sustainability, leading to further mitigation of GHG emissions.
- Devising efficient and cost-effective strategies for end use demand side management.
- Deploying appropriate technologies for both adaptation and mitigation of GHG emissions extensively as well as at an accelerated pace.
- Engineering new and innovate forms of market, regulatory, and voluntary mechanisms to promote sustainable development.
- Effecting implementation of programs through unique linkages, including civil society and local government institutions through public–private partnership.
- Welcoming international cooperation for research, development, sharing, and transfer of technologies enabled by additional funding and a global IPR regime that facilitates technology transfer to developing countries under UNFCCC.

There are eight national missions, two of which are focus on "Mitigation" and five on "Adaptation." These are National Solar Mission, National Mission for Enhanced Energy Efficiency, National Mission on Sustainable Habitat, National Water Mission, National Mission for Sustaining the Himalayan Ecosystem, National Mission for a Green India, National Mission for Sustainable Agriculture, and National Mission on Strategic Knowledge for Climate Change.

10.7.3 National Mission for a Green India

The Mission aims to respond to climate change through a combination of adaptation and mitigation measures such as enhancing carbon sinks in sustainably managed forests and other ecosystems, adapting vulnerable species/ecosystems to the changing climate, and adapting forest-dependent communities. The NAPCC states that the Green India program has already been announced, and under the program, 6 million hectares of degraded forest land would be afforested with the help of Joint Management Committees, with funds to the extent of 6000 crores being provided from the accumulated additional funds for compensatory afforestation under the decision of the Supreme Court in respect of land diverted for non-forest uses. It also suggests measures such as training on silvicultural practices for fast-growing and climate-hardy tree species; reducing fragmentation of forests by provision of corridors for species migration, both fauna and flora; revitalizing and upscaling community-based initiatives such as Joint Forest Management and Van Panchayat committees for forest management, formulation of forest fire management strategies, in situ and ex situ conservation of genetic resources, especially of threatened flora and fauna.

10.7.4 State Action Plans on Climate Change

The National Action Plan on Climate Change (NAPCC) was released on June 30, 2008 by the then Prime Minister. The NAPCC recognized the role of state and local governments in implementation of the action plan, as it was clear that climate change adaptation and mitigation challenges will only be addressed if state governments play an active role in the planning and implementing actions to achieve the objectives of the NAPCC. The subnational authorities at the state level are expected to play a key role in actively incorporating climate change considerations into day-to-day governance, adopting climate-friendly policies and programs, regulations, and investment decisions. Therefore, a process for preparation of State Action Plans on Climate Change (SAPCCs) was initiated following the announcement made by then Prime Minister in the conference of state environment ministers held on August 18, 2009. The erstwhile Ministry of Environment and Forest (MoEF) approached all state governments and union territories (UTs) to formulate SAPCCs.

The SAPCC is the first subnational exercise in climate change planning in India. After a national consultation workshop held on August 19, 2010, in New Delhi, it was decided by the MoEF that a common framework with sufficient flexibility will be provided to all state and UT governments to prepare their respective SAPCCs.

The main content of SAPCCs as per the common framework document was suggested as follows:

1. Climate profile of the state: Regional development issues vis-à-vis national priorities in the NAPCC, baseline assessments, expected future assessments, and knowledge gaps.
2. Assessment of vulnerability to climate change: Assessment of current and future vulnerability temporally and spatially.
3. GHG emissions and energy needs: Sectoral emissions and future energy needs of different sectors.
4. Climate change strategy: Identification of entry points in existing programs or schemes and cost–benefit analysis of proposed actions.
5. Climate change action plans: Identification of existing sources of funding for the plans, additional finances required, monitoring, and evaluations.

The formulation of SAPCCs can be considered an important milestone in developing decentralized domestic policies on climate change in India. It involves most development departments, the key being agriculture, water, habitat, forestry, health, and disaster management.

10.7.5 Constraints

An ideal enabling environment for an effective climate change action plan has many components such as institutional capability, awareness and understanding, evidence and research, stakeholder participation, and financing. Some of the major issues in the preparation of SAPCCs are as follows:

- Lack of dedicated and knowledgeable officials,
- Lack of coordination between states and the center,
- Limited time and resources,
- Climate change not a priority for state departments, and
- No blueprint for conducting vulnerability assessments and climate projections.

Bibliography

Adams HD, Macalady AK, Breshears DD, Allen CD, Stephenson NL, Saleska SR, Huxman TE, McDowell NG (2010a) Climate-induced tree mortality: earth system consequences. EOS Trans Am Geophys Union 91(17):153–154

Ayres MP (1993) Plant defense, herbivory, and climate change. Biotic Interactions Global change 75:75–94

Ayres MP, Lombardero MJ (2000) Assessing the consequences of global change for forest disturbance from herbivores and pathogens. Sci Total Environ 262(3):263–286

Balch JK, Massad TJ, Brando PM, Nepstad DC, Curran LM (2013) Effects of high-frequency understorey fires on woody plant regeneration in southeastern Amazonian forests. Philos Trans R Soc B 368(1619):20120157

Collins M, Knutti R, Arblaster J, Dufresne JL, Fichefet T, Friedlingstein P, Gao X, Gutowski WJ, Johns T, Krinner G, Shongwe M (2013) Long-term climate change: projections, commitments and irreversibility. In: Climate change 2013 – the physical science basis: contribution of working group I to the fifth assessment report of the Intergovernmental Panel on Climate Change. Cambridge University Press, Cambridge, pp 1029–1136

Dale VH (1997) The relationship between land-use change and climate change. Ecol Appl 7(3): 753–769

Dale VH, Joyce LA, McNulty S, Neilson RP, Ayres MP, Flannigan MD, Hanson PJ, Irland LC, Lugo AE, Peterson CJ, Simberloff D (2001) Climate change and forest disturbances: climate change can affect forests by altering the frequency, intensity, duration, and timing of fire, drought, introduced species, insect and pathogen outbreaks, hurricanes, windstorms, ice storms, or landslides. Bioscience 51(9):723–734

Day JK, Perez-Brumer A, Russell ST (2018) Safe schools? Transgender youth's school experiences and perceptions of school climate. J Youth Adolesc 47(8):1731–1742

Di Nitto D, Neukermans G, Koedam N, Defever H, Pattyn F, Kairo JG, Dahdouh-Guebas F (2014) Mangroves facing climate change: landward migration potential in response to projected scenarios of sea level rise. Biogeosciences 11(3):857–871

Doyle J (2016) Mediating climate change. Routledge

Ferry B, Bontemps JD, Blanc L, Baraloto C (2012) Is climate a stronger driver of tree growth than disturbance? A comment on Toledo et al. (2011). J Ecol 100(5):1065–1068

Füssel HM, Klein RJ (2006) Climate change vulnerability assessments: an evolution of conceptual thinking. Clim Chang 75(3):301–329

Galiano D, Kubiak BB, Marinho JR, de Freitas TRO (2013) Population dynamics of Akodon montensis and Oligoryzomys nigripes in an Araucaria forest of Southern Brazil. Mammalia 77(2):173–179

García-Valdés R, Zavala MA, Araujo MB, Purves DW (2013) Chasing a moving target: projecting climate change-induced shifts in non-equilibrial tree species distributions. J Ecol 101(2): 441–453

Gnacadja L, Lesch AK (2009) Running dry? Climate change in drylands and how to cope with it. München, Germany

Hansen AJ, Neilson RP, Dale VH, Flather CH, Iverson LR, Currie DJ, Shafer S, Cook R, Bartlein PJ (2001) Global change in forests: responses of species, communities, and biomes: interactions between climate change and land use are projected to cause large shifts in biodiversity. Bioscience 51(9):765–779

Houghton RA, House JI, Pongratz J, Van Der Werf GR, Defries RS, Hansen MC, Le Quéré C, Ramankutty N (2012) Carbon emissions from land use and land-cover change. Biogeosciences 9(12):5125–5142

IUCN, SSC (2008) Antelope Specialist Group (2020) IUCN mission to Niger for the conservation of the last wild addax and dama gazelles and the Termit and Tin Toumma National Nature Reserve: report and recommendations

Keenan RJ (2015) Climate change impacts and adaptation in forest management: a review. Ann For Sci 72(2):145–167

Keenan RJ, Reams GA, Achard F, de Freitas JV, Grainger A, Lindquist E (2015) Dynamics of global forest area: results from the FAO Global Forest Resources Assessment 2015. For Ecol Manag 352:9–20

Kelly AE, Goulden ML (2008) Rapid shifts in plant distribution with recent climate change. Proc Natl Acad Sci 105(33):11823–11826

Khaine I, Woo SY (2015) An overview of interrelationship between climate change and forests. For Sci Technol 11(1):11–18

Kodanadapani N, Cochrane MA, Sukumar R (2008) A comparative analysis of spatial, temporal, and ecological characteristics of forest fires in seasonally dry tropical ecosystems in the Western Ghats. India For Ecol Manag 256:607

Kumar V (2018) Coping with climate change: an analysis of India's State Action Plans on Climate Change. Centre for Science and Environment, New Delhi

Lenoir J, Gégout JC, Guisan A, Vittoz P, Wohlgemuth T, Zimmermann NE, Dullinger S, Pauli H, Willner W, Svenning JC (2010) Going against the flow: potential mechanisms for unexpected downslope range shifts in a warming climate. Ecography 33(2):295–303

Levina E and Tirpak D (2006) Adaptation to climate change. Key terms

Liverman DM (1990) Drought impacts in Mexico: climate, agriculture, technology, and land tenure in Sonora and Puebla. Ann Assoc Am Geogr 80(1):49–72

Lucier A, Ayres M, Karnosky D, Thompson I, Loehle C, Percy K, Sohngen B (2009) Forest responses and vulnerabilities to recent climate change, vol 22. IUFRO World Series, Helsinki, pp 29–52

Mátyás C (2010) Forests and climate change in Eastern Europe and Central Asia, vol 8. FAO, Rome

McEvoy D, Fünfgeld H, Bosomworth K (2013) Resilience and climate change adaptation: the importance of framing. Plan Pract Res 28(3):280–293

Opler PA, Baker HG, Frankie GW (1980) Plant reproductive characteristics during secondary succession in neotropical lowland forest ecosystems. Biotropica 12:40–46

Parada T, Lusk CH (2011) Patterns of tree seedling mortality in a temperate-mediterranean transition zone forest in Chile. Gayana Botanica 68(2):236–243

Pearce W, Holmberg K, Hellsten I, Nerlich B (2014) Climate change on Twitter: topics, communities and conversations about the 2013 IPCC Working Group 1 report. PLoS One 9(4):e94785

Savely HE (1939) Ecological relations of certain animals in dead pine and oak logs. Ecol Monogr 9(3):321–385

Seppälä R (2009) Adaptation of forests and people to climate change: a global assessment report, vol 22. IUFRO, Helsinki

Seufert P (2013) The FAO voluntary guidelines on the responsible governance of tenure of land, fisheries and forests. Globalizations 10(1):181–186

Smith WK, Germino MJ, Johnson DM, Reinhardt K (2009) The altitude of alpine treeline: a bellwether of climate change effects. Bot Rev 75(2):163–190

Stone WE, Wolfe ML (1996) Response of understory vegetation to variable tree mortality following a mountain pine beetle epidemic in lodgepole pine stands in northern Utah. Vegetatio 122(1): 1–12

Suresh MVJJ, Reddy KS, Kolar AK (2010) 4-E (Energy, Exergy, Environment, and Economic) analysis of solar thermal aided coal-fired power plants. Energy Sustain Dev 14(4):267–279

Suzuki SN, Tsunoda T, Nishimura N, Morimoto J, Suzuki JI (2019) Dead wood offsets the reduced live wood carbon stock in forests over 50 years after a stand-replacing wind disturbance. For Ecol Manag 432:94–101

Thompson I, Mackey B, McNulty S, Mosseler A (2009) Forest resilience, biodiversity, and climate change. Secretariat of the Convention on Biological Diversity, Montreal Technical Series no. 43.1–67. (Vol. 43, pp. 1-67)

Vignola R, Locatelli B, Martinez C, Imbach P (2009) Ecosystem-based adaptation to climate change: what role for policy-makers, society and scientists? Mitig Adapt Strateg Glob Chang 14(8):691–696

Whelan RJ (1995) The ecology of fire. Cambridge University Press, Cambridge

White GC, Bennetts RE (1996) Analysis of frequency count data using the negative binomial distribution. Ecology 77(8):2549–2557